The Discursive Interview

Carsten G. Ullrich

The Discursive Interview

Method and Methodological Foundation

Springer

Carsten G. Ullrich
Fakultät für Bildungswissenschaften
Universität Duisburg-Essen
Essen, Germany

ISBN 978-3-658-38476-0 ISBN 978-3-658-38477-7 (eBook)
https://doi.org/10.1007/978-3-658-38477-7

This book is a translation of the original German edition „Das Diskursive Interview" by Ullrich, Carsten G., published by Springer Fachmedien Wiesbaden GmbH in 2020. The translation was done with the help of artificial intelligence (machine translation by the service DeepL.com). A subsequent human revision was done primarily in terms of content, so that the book will read stylistically differently from a conventional translation. Springer Nature works continuously to further the development of tools for the production of books and on the related technologies to support the authors.

This Springer imprint is published by the registered company Springer Fachmedien Wiesbaden GmbH, part of Springer Nature.
The registered company address is: Abraham-Lincoln-Str. 46, 65189 Wiesbaden, Germany

Contents

List of Figures

List of Tables

Introduction

This textbook introduces the research method of the Discursive Interview. Despite its relative brevity, it is intended to provide a comprehensive insight into the methodology and practice of this interpretative method. The explanations therefore cover both the practical procedure of conducting and interpreting Discursive Interviews and the methodological considerations underlying the Discursive Interview.

The explanations also react to numerous inquiries, objections and needs for clarification with which I have been confronted again and again since the first presentation of the Discursive Interview (Ullrich 1999), and in this respect also aim at clarifying many such questions more permanently. Nevertheless, with regard to the Discursive Interview, there are of course further needs for clarification and open questions, which will be made clear here at the given points.

The Discursive Interview was developed for the recording and reconstruction of social patterns of interpretation and is therefore primarily understood as a method for analysing patterns of interpretation, based on a sociological understanding of social patterns of interpretation. The explanations and justifications of the individual methodological steps, however, often require more general explanations of fundamental questions of qualitative social research or of interview methods. This can sometimes be tedious for readers with prior knowledge and experience of qualitative methods, but at the same time it opens up the possibility of being of interest to other fields of research. Thus, in view of the many unresolved issues and "white spots" in qualitative interview methodology, I at least have the hope that some of the considerations presented here can also contribute to improving the understanding of methodological issues related to qualitative interviews in general.

Before the Discursive Interview can be explained, it is necessary to briefly clarify what is understood here as a social interpretive pattern and what interpretive pattern analysis consists of (Chap. 1). A theoretical discussion of the interpretive pattern concept is not the central concern of this book, but it is unavoidable in an operational sense in order to clarify how the Discursive Interview responds to the characteristics of social interpretive patterns and attempts to use their communicative functions.

On this basis, the theoretical and methodological basic assumptions of the Discursive Interview are presented in detail in the second chapter. In doing so, it becomes clear that the Discursive Interview sees itself as an integrated research design that encompasses all research steps. In addition to the general premises of the sociology of knowledge, this chapter primarily discusses the interactional setting of the research interview and its implications for data collection. Another aspect that is central here are the logics and structures of different stimulus and question forms as well as their effects on the answers of the interviewees and the text types that emerge in Discursive Interviews.

Chapters 3 and 4 then deal in detail with the two main phases of qualitative-empirical research, data collection and data interpretation. In the third chapter, the different types of questions and stimuli used in discursive interviews are central. In this context, it is also explained which special question forms for eliciting interpretative patterns are important for the Discursive Interview and how they can be used. In addition, this chapter explains sampling and recruitment and discusses problems in conducting Discursive Interviews (including the use of guidelines, research ethics issues).

For the data interpretation (Chap. 4), it is crucial how social patterns of interpretation can be reconstructed from the obtained text material. For this purpose, the Discursive Interview provides a contrasting interpretation strategy, which is justified and explained in detail. In this strategy, social patterns of interpretation are successively extracted from individual patterns of interpretation or derivations through systematic and repeated comparisons. Other topics discussed in this section include issues of transcription, the appropriate inclusion of contexts, and forms and steps of type formation.

Finally, the fifth and last chapter deals with aspects of quality assurance. Here, the main focus is on how to ensure and clarify the credibility of research results obtained with Discursive Interviews and how to increase their reach (scope).

Collective Representations and Their Analysis

1

The discursive interview method was developed as an instrument for eliciting and reconstructing collective representations. In order to understand the Discursive Interview, it is therefore necessary to clarify what collective representations are and why the reconstruction of collective representations is important in the social sciences. Above all, however, the methodological procedures of the Discursive Interview are based on a specific understanding of collective representations; they cannot be understood without an idea of what (collective) representations are and how they function.

This first chapter therefore first deals with the concept pf collective representations and their analysis. The first section attempts to extract from the thicket of different approaches and definitions a working definition of social or collective representations that is sufficient for the understanding of the Discursive Interview (Sect. 1.1). After this necessary preliminary terminological work, the second section deals with the analysis of collective representations, that is, with the spectrum of empirical strategies for reconstructing social (collective) representations (Sect. 1.2). The last section then presents some research examples in synoptic form (Sect. 1.3).

1.1 Collective Representations

In this section, we will first clarify in general terms what is meant by the term collective representation for our purposes in the further course. In other words, the question is how the term collective representation can be meaningfully defined and how the this concept differs from other concepts (Sect. 1.1.1). A fundamental

C. G. Ullrich, *The Discursive Interview*,
https://doi.org/10.1007/978-3-658-38477-7_1

exegesis of the term or comprehensive classificatory delimitations will not be attempted; they will only be carried out to the extent necessary to gain a basic understanding sufficient to explain the Discursive Interview. In doing so, it should also be made clear what the social scientific interest in a reconstruction of collective representations is based on.

After these clarifications, the central properties or functions of collective representations are briefly presented (Sect. 1.1.2). In the later remarks on the methodological procedure (esp. Chaps. 3 and 4), repeated reference is made to some of these properties. Finally, the distinction between collective representations, individual representations and (individual) derivations, which is important for the discursive interview, is explained (Sect. 1.1.3).

1.1.1 What Are Collective Representations?

The closer definition of what is referred to in the following as collective representations is important for two reasons: On the one hand, the research interest in (collective) representations explains the general methodological interest in the question of how collective representations can be captured and reconstructed. The social science-based interest in collective representations is the central prerequisite for the discursive interview as a research method. Furthermore, the concrete procedures of the Discursive Interview are explained by (my) specific understanding of the nature and mode of action of collective representations. As will become even clearer in Sect. 1.2: To the extent that (collective) representations are defined differently (than suggested here) or even that other features are seen as paramount, one may also tend towards a different methodological procedure than the Discursive Interview.

Thus, the aim is not a comprehensive conceptual exegesis or even a theory of collective representations (or even preliminary considerations for such a theory), but rather an explanation of a general understanding about social reality and how it can be approached methodologically. The explanations in this section therefore do not claim to be original and are not intended to open up new aspects for a theory of collective representations. Rather, the aim is to try to clarify the (as I hope: relatively consensual) core ideas of the concept of collective representations.

The term "collective representation" (French: "représentation collective") was originally coined by Emile Durkheim (1981 [1912]). But there does not exist a common terminology. Thus, many alternative or competing terms are used, mostly without an elaborate definition.[1]

[1] In German the term "soziale Deutungsmuster" is predominantly used. It treated here synonymous to (collective) representations.

Definitions and Approaches to Interpretation

The term collective representation is used quite differently and often in rather loose terms. It is rarely clearly defined, nor is the collective representations concept well demarcated from alternative and competing concepts. On the one hand, this is most certainly due to the fact that there is no unified theoretical concept or theory of collective representations.[2] Moreover, the introduction of the concept of (collective) representations into (German-language) theory and into qualitative social research came from different sides and therefore also from different methodological and theoretical positions and with different objectives. In no case, however, has it acquired a central theoretical status.[3]

Secondly, this broad but fuzzy introduction is accompanied by a relatively high popularity of the concept of (collective) representations and a correspondingly unspecific use, with (collective) representations usually referring to forms of knowledge that are not further defined. Finally, there are a number of alternative or similar terms (such as metaphors, orientation patterns, cognitive maps, stereotypes,

[2] The lack of a theory of collective representations (soziale Deutungsmuster) has been lamented since the beginning of the use of the concept of representations (for attempts at a stronger theoretical foundation and classification, however, see e.g. Keller 2014; Konderding 2008; Plaß and Schetsche 2001). Whether there is a need for such a theory and whether it would be possible at all, however, is to be doubted: there is no need because "representation" is a concept and not a system of statements. As a concept, it is based on (pre-)theoretical premises or axioms whose viability must be proven in their implementation in theory and research. Moreover, like many other basic sociological concepts (e.g. role or interaction), the representation concept has developed in different theoretical traditions. Therefore, an "isolated", independent theory of collective representations is not conceivable, but at best different representation theories. In contrast, further developments of explanatory approaches (especially in the sociology of knowledge and communication theory) that focus on the relevance of different forms of knowledge for action could certainly provide more insight into the functioning of collective representations.

[3] With Arnold (1983) we can distinguish three essential theoretical impulses for the interest in collective interpretations: These are the sociology of knowledge, especially in the tradition of Schütze ((1983), Schütz and Luckmann (1979) and Berger and Luckmann (1990), the so-called worker consciousness research (Popitz et al. 1957; Neuendorff and Sabel 1978; Thomssen 1980), and the objective hermeneutics of Oevermann (1973, 2001; Oevermann et al. 1979), which starts from structuralist ideas. The historical-genetic perspective (e.g. Honegger 1978, 2001; Schütze 1986) should also be added here. For the definition and theoretical embedding of the representation concept, see, among others, Keller (2014), Lüders (1991), Lüders and Meuser (1997), Meuser and Sackmann (1992), Müller (2013), Plaß and Schetsche (2001), and Schetsche (2000).

scripts, interpretive frameworks/frames)[4] that overlap in meaning with that of collective representations – or at least appear to overlap, for these concepts are also rarely precisely defined and are used as frequently as they are blurred.

Comparative analyses of these concepts or even a typology of action-guiding forms of knowledge would be necessary for a clearer and well-founded differentiation of action-guiding forms of knowledge and for structuring this field. However, this is still a long way off. This circumstance, which is certainly not uncommon in the field of social sciences, is probably due, among other things, to different disciplinary backgrounds – concepts of forms of knowledge are particularly widespread in sociology, psychology and communication sciences – as well as to the different theoretical embeddings and points of reference.

There is therefore no "binding" or consensual definition of (collective) representations. A comparison of different definitions would therefore reveal that different characteristics and functional contexts are named or accentuated by the researchers and theorists. Nevertheless, clear overlaps can be identified, so that a definitional intersection can be identified that is shared, if not by all, then at least by most of those who refer to the concept of collective representations.

In my opinion, a definition that is very close to this intersection is still that of Arnold (1983, p. 894). According to this, collective representations (resp. Deutungsmuster) "can be described as the more or less time-stable and in a certain sense stereotypical views and interpretations of members of a social group which they have developed in relation to their everyday areas of action and interaction over the course of their lives. Specifically, these (collective) representations constitute an orientation and justification potential of everyday knowledge stocks in the form of basic, rather latent situational, relational, and self-definitions in which the individual presents his identity and maintains his capacity to act."

[4] The sometimes very different theoretical references of these and other terms cannot be addressed here. Unless these terms are simply used non-specifically, it is usually not possible to clarify differences between representations and these concepts in principle and in general terms due to the often contradictory theoretical frameworks. Already here, however, it can be stated that differences often exist, or at least can exist, and that (collective) representations cannot simply be considered synonymous with "script", "metaphor", and others.

This characterization can serve as a working definition (see also Sect. 1.1.2).[5] Accordingly, it is essential for collective representations that they

- are collectively shared ("individual" representations are therefore never independent of social ones),
- can be regarded as everyday models or theories,
- have to prove themselves in practical action, i.e. have to be "successful", but not "true",
- have a minimum degree of complexity (i.e. cannot be reduced to individual, especially cognitive and normative elements or functions) and
- are communicated and used to legitimize actions.

In the following, therefore, a broad understanding of collective representations is assumed. According to this, collective representations are composed of different units and can, for example, contain normative and evaluative elements in addition to cognitive ones. In addition, collective representations usually refer to "middle-level" forms of knowledge that are more complex and abstract than simple type terms, but on the other hand more "small-scale" than, for example, ideologies or discourses.

---

[5] Nevertheless, at least three differences to my understanding of collective representations should be named here:

(1) Thus, in my opinion, there is no reason why collective representations should be limited to social groups *by definition*. The social spread and distribution of representations is an empirical question. In addition to class-, milieu- or group-limited representations, individual, familial or cultural and historical ones can therefore also exist.
(2) An understanding of representations as "situational, relational and self-definitions" and as "stereotypical views and interpretations" is broad, but nevertheless a narrowing that is not necessary. Why should there not, for example, also be representations in relation to objects, social groups and categories, historical processes, and so on?
(3) The fact that collective representations form "an orientation and justification potential (…) *in the form of* basic (…) situation, relationship and self-definitions" (emphasis mine) suggests that representations *are* definitions of a situation. In contrast, representations are assumed here to be *applied* in situational definitions (or more generally: to problems of reference), that is, to coincide with situational definitions only exceptionally.

Collective Representations of Poverty (an Example)

"Poverty" can be seen as an example of an extremely complex phenomenon, for which there are a large number of competing, socially unequally distributed and historically and culturally varying collective representations. Depending on the type of collective representation used, poverty is defined differently and is composed of different individual aspects. (collective) representations of poverty therefore include[6]:

- Poverty definitions, including: Types of poverty (income poverty, educational poverty, etc.), poverty measurement (relative vs. absolute poverty), social standards and thresholds (When is someone poor? Are e.g. recipients of unemployment benefit, visitors of food banks etc. poor?), cultural/historical relativity of poverty (e.g. poverty in developing countries).
- Causes of poverty, including: Inevitability of poverty, poverty as a consequence of political or economic aberrations, poverty as a consequence of poor people's misbehaviour.
- Which social groups are (particularly) often poor (young, old, women, children, single parents, migrants, unemployed?)? How high is the risk of poverty (including one's own)?
- Consequences of poverty, including: for "the poor" (such as material hardship, perpetuation, stigmatisation), for the non-poor (including fear of crime, social conscience), for society (e.g. social conflicts, exclusion, repression or problematisation of forms of poverty), for politics (including socio-political combating of poverty, making poverty invisible)
- Attitudes to poverty, including: Fear of poverty, positive perceptions of poverty (voluntary poverty).

The aspects listed here, which are by no means exhaustive, already make it clear that collective representations related to poverty (must) lead to reductions in complexity in different ways. Depending on which aspects are brought to the fore, a distinction can be made, for example, between trivialising ("nobody is really poor in this country"), dramatising ("poverty is on the increase"), victimising ("it is mostly the poor's own fault"), socialising ("poverty must be prevented politically") and fatalistic ("poverty can affect anyone") representations.

[6] Cf. inter alia: Gans (1992); Geremek (1991); Groenemeyer (1999); Leibfried and Voges (1992); van Oorschot and Loek (2000); De Swaan (1993).

Collective representations can thus be described as the ideas shared by a collective about comparatively complex phenomena. These include poverty, unemployment, democracy, ethnicity, war, nationality, social problems, social inequality or bisexuality (but also less "conspicuous" things like family, work, school, etc.). Therefore, it is also too simplistic to assume that collective representations provide actors with clear options for action (if not instructions for action). Rather, they usually contain multi-layered and by no means necessarily consistent interpretations of a phenomenon (and corresponding options for action).

The Sociology of Knowledge Approach to Collective Representations

An interest in collective representations can develop from different methodological directions. However, particularly frequent in the field of collective representation analysis are positions from the sociology of knowledge. This is not surprising, considering that the sociology of knowledge is interested in the nature and effect of different forms of knowledge. The Discursive Interview is also based on a sociological of knowledge understanding of collective representations.

The interest in reconstructing collective representations (or, more generally, social meaning) derives from three main axiomatic assumptions:

1. Assumption of relative autonomy and the action-generating effect:
   The first is that knowledge of socially widespread representations is sociologically important, i.e. it is significant for the explanation of socially typical actions (patterns of action) and phenomena. Although they may have complex interactions with social-structural factors, they are not their "epiphenomena", i.e. they cannot be derived exclusively from them.
2. Assumption of collectivity of representations:
   Secondly, the interest in the analysis of collective representations arises from the conviction that individual perceptions and interpretations of the social and physical world, as well as action orientations and actual actions based on these perceptions and interpretations, are dependent on collective representations and legitimations (more generally: collective stocks of knowledge), i.e. cannot be imagined without them. The core assumption, then, is that collective representations are crucial for actors' ability to act, and therefore also for explanations of action. To put it differently: If we want to explain actions and patterns of action, we need to know the collective representations on which actors base their actions.

3. Assumption of latency of collective representations and a need for reconstruction:
   The third premise states that collective representations are not "obvious", not immediately recognizable or observable. Rather, they form a latent basis for decisions, actions, judgments, etc., i.e. for directly observable phenomena. The collective representations underlying these manifestations must and can only be reconstructed from these observable evidences in a comprehensive process of analysis.

A sociology of knowledge concept of collective representations must be distinguished from (radical) *constructivist* positions. Thus, sociological analyses of collective representations assume that collective representations stand in a more or less clear functional relation to an "objective situation" (for the acting person applying a collective representation) or problem of reference arising from a situation.[7] In this context, it must be assumed that there is a close interdependent relationship between the collective representations and the problem of reference: What is perceived by the agents as a situation or problem of action depends, on the one hand, considerably on the representations socially available to them. On the other hand, situational constellations confront agents with problems of action that are "objective" for them.[8] They are "social facts" in the Durkheimian sense (Durkheim 1984 [1895]).

Regardless of the role that (collective) representations play in defining the situation, successful and stable representations will always have a certain situational adequacy and will "work" in accomplishing the tasks of action. However, since an

[7] If, in the following, that to which representations refer is called a problem of reference, this is done in the absence of a more suitable, above all a theoretically more substantial term, but also in connection with a widespread use of language in the context of the analysis of collective representations. The designation as a problem of reference, however, is in no way intended to limit representations to the realm of "social problems" or individual problems of decision and action. What is "problematic" here (at least from the perspective of the user of representations) is solely the interpretation of a phenomenon, which then leads to the application of collective representations. The fact that problems of interpretation then often lead to individual problems of action and that collective representations then enable action is a frequently observed consequence, but not a necessary condition for the existence of representations.

[8] In contrast to postmodernisms and radical constructivisms so widespread today, Oevermann (1973, p. 4) was already refreshingly clear about this: "Of course, objective problems of action always already enter the field of action of the subject as culturally interpreted problems, that is, as problems interpreted in terms of representations. In this respect, representations are always at the beginning of a sociological causal analysis. On the other hand, collective representations cannot be explained without referring back to objective problems of social action to which they respond."

optimal fit between the problem of reference and the collective representations – whether the situation definition completely determines the (collective) representations or, conversely, is completely determined by them – is at best a theoretical exceptionality, a permanent process of reciprocal adjustments of representations and the problem of reference must be assumed. Representations are thus always exposed to the danger of negative tests of experience.

Another important consequence of a concept of (collective) representations from the sociology of knowledge is that it eludes simple localization at the level of action or structure. This enables an explanation of patterns of action without deterministic or voluntaristic reductions. Thus, representations are understood as social phenomena that cannot be attributed to individual characteristics. Accordingly, actions are primarily understood as a consequence of (collective) representations that guide action. At the same time, however, it is precisely the individual transformational achievements that are focused on. Collective representations increase the action competence of actors by generating individual action orientations in the first place, but at the same time only "legitimated" ones, i.e. those that are within the framework of the dominant discourse. Action orientations and action decisions can thus be understood and analysed as products of social contextual conditions.[9]

An analysis of collective representations in the sociology of knowledge thus focuses on the *constitutional conditions of action orientations* and places the reconstruction of action-guiding collective representations at the center of research interest. The theoretical and methodological challenge is to show which representations actors refer to and how they transform them into situation definitions and action orientations.

1.1.2 Properties of Collective Representations

It is generally assumed that collective representations are an essential prerequisite for the ability of individual actors to act. They represent a central part of the opportunity structure of social action (cf. Arnold 1983) and thereby define the spec-

[9] The fact that this theoretical potential of collective representations is still not sufficiently exploited is perhaps also due to the widespread focus in the social sciences on Bourdieu's concept of habitus (Bourdieu 1976), which – despite early emphasis on parallels and complementary orientations (Matthiesen 1989, among others) – has led to an adherence to the structuralist bias of the habitus concept. As late as 2001, Oevermann also emphasizes the closeness of his understanding of collective representations (Deutungsmuster) to the habitus concept (Oevermann 2001, p. 46).

trum of possible forms of action: Only insofar as we can fall back on collective representations can we react in situations and to problems of action; and at the same time, the socially available representations always open up only a specific selection of "socially legitimate" options for action from the spectrum of conceivable ways of reacting. Collective representations are thus at the same time *action-enabling as well as action-controlling.*

Collective representations generate individual agency not only by offering simple heuristics, but also by normative, evaluative and practical elements. Collective representations are thus relatively complex. They "tell" an agent not only what the case is, but also how a given action situation is to be evaluated (normatively, aesthetically) and which forms of reaction are possible or meaningful.[10]

However, generating agency by collective representations is only possible if the basically infinite possibility of situation definitions and problem perceptions is limited. Even relatively complex representations become enabling in that they reduce the complexity of the situation and also thereby offer a coherent (seemingly consistent) definition of a situation. They simplify complex contexts and make conflicting interpretations, norms and values compatible. *Complexity reduction* and *"consistency fiction"* are mechanisms for establishing the ability to act and should not lead to the assumption that collective representations are truly consistent or reduce complexity. As Oevermann (1973, p. 24) already emphasized "collective representations are never a completely closed and self-contradictory system of interpretations".

Other properties often attributed to collective representations concern their degree of consciousness, their logical structure, and the importance of proving experience.

For enabling action, collective representations usually remain latent in the action process (assumption of a relative *latency*). Actors usually have to quickly grasp the situations in which they find themselves and to which they have to react. Collective representations are called up "in a flash" and applied to the situation at hand; they are thereby "only to a limited extent reflexively available" (Meuser and Sackmann 1992, p. 19). In a large part of social interactions there is hardly any opportunity for longer reflections – even if it is only the search for a suitable representation. For it is precisely through the automatic "retrieval" of collective representations that actors are able to react both quickly and "correctly" (in relation to the situation). Conversely, this means: If actors encounter situations for which they

[10] It is primarily this complexity that distinguishes collective representations from almost all "competing" concepts (see above). This applies to all primarily cognitive concepts (cognitive maps, scripts, frames, metaphors), but also, for example, to normative ones (attitudes, values).

do not have any suitable collective representation, they can only react slowly because they first have to analyze the situation.

This does not mean, however, that (collective) representations are not reflexively available to the agents, i.e. that they remain unconscious. On the contrary: individual representations can be made the object of self-reflection in a targeted manner (even if not in the course of action). Often enough, agents are even forced to do so: on the one hand, when their individual representations fail; on the other hand, when they are confronted with other, possibly contradictory representations.

Another feature of collective representations is already implicit in the effects described: collective representations are themselves *neither "logical" nor consistent*. They are successful to the extent that they enable action or offer the actors plausible world views. To this end, it is virtually necessary that they simplify social reality. Users of collective representations must therefore have a certain tolerance for inconsistency, i.e. they must apply (collective) representations even if they are obviously contradictory.

This applies not only to single representations, but even more so to the totality of an actor's collective representations. In principle, it can be assumed that there is a "primordial" conglomerate of hierarchical organized representations that differ, among other things, in their degree of generalization and thus in their scope of application, and that are often related to each other. In many situations, therefore, actors will often be able to fall back on different, competing and possibly mutually exclusive (collective) representations, but will also have to endure and balance the cognitive dissonances.

Collective representations are *dependent on experience*, but at the same time also relatively *resistant to experience:* On the one hand, representations cannot exist in the long term if they are "unreal" and do not enable the user to act appropriately in a situation. On the other hand, collective representations that have proven themselves in earlier situations are not corrected at every minor irritation. This will only happen if collective representations prove to be permanently inadequate.

The application of wrong, because situationally inappropriate representations can have considerable, in extreme cases fatal disadvantages. Collective representations are therefore reflected upon and revised by the users at the latest when the consequences of action are unsatisfactory and can be traced back to incorrect definitions of the situation. In most cases these will only be minor corrections, additions, differentiations and readjustments. In contrast, a final discarding of a previously proven collective representation is likely to be a rare (and possibly crisis-ridden) exception. The ability to make such adjustments to the set of individual representations must be regarded as a fundamental prerequisite for the ability to act. Permanent adherence to obviously inadequate representations is usually socially pathologised.[11]

Two other closely related properties of social (collective) representations are central to the Discursive Interview method: collectivity and the legitimizing function of collective representations.

1. Social representations are best understood as part of a collective store of knowledge. This means that collective representations are a *social, emergent phenomenon:* They always manifest themselves in individual consciousness only partially and in specific individual-idiosyncratic transformations. Collective representations are thus socially shared and communicated – in a social and cultural space that cannot be further defined here. They are learned by the individual (initially in the socialization process) and must be repeatedly "communicatively validated" (i.e. confirmed by members of one's own group or culture). (If this is not the case, i.e. if an actor deviates too far from the socially accepted representations, there is also a danger of pathologisation).

Conversely, this means: Actions must be traced back to socially accepted (or assumed to be socially shared) representations, both for external *legitimation* and for the actors themselves. Actors can only present their actions as "meaningful" to the extent that they correspond to interpretations shared (or at least known and tolerated) by other members of the common space of experience. Due to their collective character, representations thus also have normative validity (cf. Meuser and Sackmann 1992, p. 19).

1.1.3 Collective Representations, Individual Representations and Derivations

So far, the concept of collective representations has been used rather unspecifically. For our purposes, however, three types or levels of representations should be conceptually distinguished:

[11] The scope of social reactions is wide, ranging from irritation in interactive situations to culturally accepted or tolerated forms of deviance and psychotherapeutic interventions. The social reactions are themselves guided by collective representations. As the social treatment of religious interpretations in particular shows, these can also – and against all empirical evidence – be imposed on individual actors.

1. These are first of all *collective representations* which, as part of a collective stock of knowledge, are the "property" of this collectivity. They form the horizon of legitimate and possible interpretations. They are socially shared in the sense that it can be assumed that, as part of a collective consciousness, they are the basis for successful interactions and communications in an identifiable social space. This set of socially available (collective) representations is amorphous and inconsistent. The collective store of knowledge also provides competing and contradictory representations. It can be assumed that no member of an interpretive collective knows or even applies all available collective representations. This applies to the entire spectrum of social representations, to the competing (collective) representations of a reference problem as well as to individual representations. *Individual representations* therefore (a) always represent a selection or subset of the available collective representations and (b) are necessarily individual variations of collective representations. No single member therefore represents (complete) collective representations, but always only individual adaptations and that means always partial, if not alternated, versions of collective representations.
2. These individual adaptations of collective representations orient the situational actions of actors. A distinction must be made between collective representations that are used to communicate and legitimize actions. These will be referred to here as *derivations.*[12] Derivations are also applications of collective representations. Although derivations cannot be equated with collective representations, they are an expression of what is assumed to be socially shared. Derivations are also identical with the action-guiding individual representations only in borderline cases, but will usually have a larger intersection with them. In principle, however, the derivations used in interactions to justify action are likely to deviate more or less from the individual representations – not least because derivations have to be explicated to a much greater extent. Finally, the collective representations are repeatedly confirmed, reproduced and changed through their use as derivations (see Fig. 1.1).

[12]The concept of derivation goes back to Pareto (1955 [1916]) and, together with its structural counterpart of the residuum, forms the core of Pareto's sociology of knowledge (cf. also Eisermann 1962, p. 170 ff.). The concept of derivation used here differs from Pareto's in that it is used here to denote only "representations used in justifications of action" (and not all forms of "rationalizations of action"). At the same time, the ideological assumption that is at least implicit in Pareto is dispensed with.

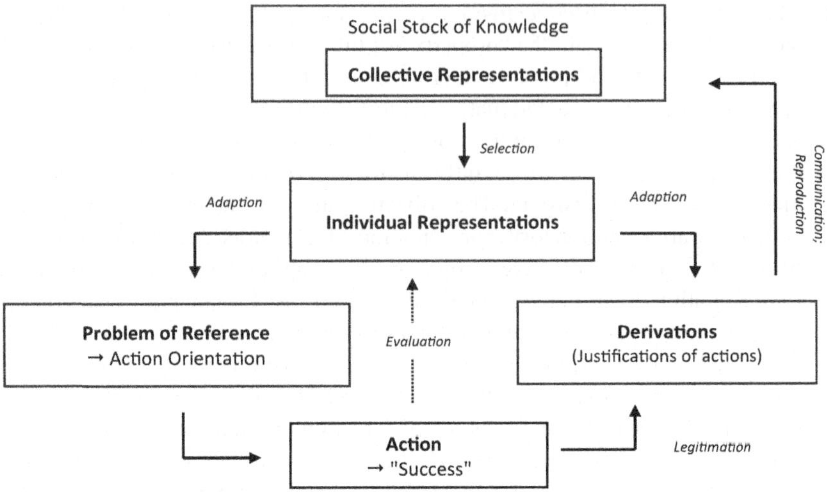

Fig. 1.1 Collective representations, individual representations and derivations

As will be explained in the following chapters (especially Chaps. 2 and 4), derivations are of decisive importance for the method of the discursive interview, which makes use of the communicability and simultaneous communicability necessity of collective representations.

1.2 The Analysis of Collective Representations

Collective representations analysis is not a method, but a research area or approach that establishes a specific research interest in (collective) representations and derives methodological guidelines from this interest.

Empirical access to collective representations can take place in different, though not arbitrary, ways. However, collective representations analysis is also more than just a research field or subject. Not every interest in collective representations leads to a collective representations analysis approach. If, for example, only the socio-spatial distribution of collective representations is to be "mapped", a reconstructive analysis of collective representations is not necessary.

In contrast, research forms for which the reconstruction of collective representations is central are to be described here as collective representations analysis. Here, too, different methods are used, but they all share this basic objective. Collective representations analysis can thus be distinguished from approaches and research interests in which collective representations play a role or even represent a central element of the analysis, but which do not see their research goal in reconstructing collective representations.[13]

Collective representations analysis research must therefore accomplish two things: "find" collective representations and, because they are not immediately visible, reconstruct them (and thus make them visible). To do this, the first step is to obtain *material that contains (collective) representations.* This material can be very different: Often "natural" documents (e.g. newspaper articles) or texts generated in the research process (e.g. interview transcripts) will be used; but also image and video material (especially multimedia material) can be used for an analysis of collective representations.

Once this material has been obtained, the reconstruction of collective representations can begin. Reconstruction is necessary because collective representations, as a socially shared form of knowledge, cannot simply be found, established or measured. As we have seen (see Sect. 1.1.1), collective representations are to a certain extent "hidden" in individual representations and derivations.

What we find in data form – primarily in texts – are first of all individual *derivations.* But of course not every text contains derivations or even consists only of such. Therefore, the data material must first be analysed to see whether it contains derivations. If this is the case, these individual derivations manifesting themselves in texts can be opened up interpretatively.

Thus, it cannot be assumed that the individual derivations can be asked directly or recorded in any other way, or even that the users themselves identify their derivations. The relative latency of the representations, which will hardly be present to the users, especially in the case of verbal utterances, speaks against this. In addition, the relatively high complexity of derivations requires the reconstructive development of the collective representations or derivation-specific context of meaning.[14]

---

[13] Of course, this does not exclude the possibility that research on representations and other research directions concerned with representations can fruitfully complement each other. Thus, for the analysis of collective representations and their forms of impact, "divisions of labour" can be imagined very well, especially with historical and sociological discourse analysis and conversation analysis. Discourse analysis is in any case often explicitly concerned with the processes of the emergence and enforcement of collective representations (e.g. Honegger 1978; Schütze 1986) or considers representation analyses as part of a discourse reconstruction (Keller 2014). Conversation-analytical work, on the other hand, could shed light in particular on the communicative function and practice of representations (legitimation, validation).

[14] For this reason, it has also proven difficult to at least prove representations with standardized methods (cf. Lüdemann 1992; Ullrich 2008). The collective representations approach is obviously not very compatible with the variable and subsumption logic of standardized research.

Fig. 1.2 Steps of the collective representations analysis

It is now of fundamental importance for a collective representations analytical perspective that *collective representations* are never directly accessible, but can only be captured via the detour of individual applications of the corresponding collective representations. These often idiosyncratic adaptations must be analysed, condensed and cleansed of user-specific characteristics in such a way that they reveal consistent collective representations. The reconstruction of collective representations is thus carried out on the basis of the individual derivations already uncovered or in a parallel process of reconstructing derivations and (collective) representations (cf. also Sect. 4.2).

Often an analysis of collective representations will only be considered complete when all collective representations of a problem of reference have been reconstructed. This also includes internal differentiations of the usually rather small number of collective representations of a reference problem. This is often attempted in the form of a *typology*. However, this presupposes that the necessary prerequisites for this are already present in the data material during the overall planning and data collection (cf. also Sect. 4.4).

Altogether, four steps or stages are to be distinguished (see Fig. 1.2): the acquisition of data material from which collective representations can be reconstructed, the reconstruction of the derivations, the reconstruction of collective representations, and finally the development of the field of empirically observable or reconstructable collective representations. Only the first three steps are indispensable components of an analysis of collective representations. Whether the entire field of collective representations of a problem of reference should also be grasped and reconstructed depends on the respective research question.

1.2.1 The Eliciting of Individual Derivations

Assuming that an analysis of collective representations must primarily draw on texts as data material, a distinction can first be made between two basic ways of collecting data (or "obtaining material") and two basic types of data: The data material can either be specifically generated in the research process or exist in "natural" data; and it can be in oral or written form. Altogether, this results in four basic strategies for obtaining material suitable for analyses of collective representations:

1. Often, *verbal data generated in the research process* are used for collective representations analyses. In almost all cases, this takes place in the form of qualitative interviews or group discussions and on the basis of transcriptions of verbal utterances. The advantage of a targeted generation of data material in the research process that contains derivations or collective representations is mainly to be seen in the fact that the researchers can provide for rich material through strategies directed at the elicitation of derivations. A disadvantage of this approach could be seen in the reactivity of interviews and group discussions. This is especially a problem if the actual collective representations guiding action are to be reconstructed, which, as explained, do not have to be identical with the derivations revealed in conversations.
2. However, an analysis of collective representations is not dependent on oral data. Thus, derivations and collective representations can also be reconstructed from *written texts generated in the research process*. The prerequisite for this is that research subjects are successfully encouraged to write corresponding texts through stimuli and incentives. This can take place primarily in the form of a written qualitative questioning (cf. Schiek 2014) or through other stimuli for written text production (classic: essay or diary). Qualitative-written forms of data collection are associated with many difficulties, among which the motivation problem[15] might be the most important. However, the possibility of using new media has recently made this form of data collection much more attractive. For example, online interviews and online group discussions (chats, message boards) in particular can be used quite well to produce written data that also seem suitable for analyses of collective representations (cf. among others Früh 2000; Ullrich and Schiek 2014).
3. The third way of "obtaining material" for an analysis of collective representations is the recording or collection of *"natural" oral data* (especially recorded conversations). These enable the capturing of derivations uninfluenced by the effects of data collection (e.g. interviews). Overall, however, natural conversation data do not appear to be very suitable as a primary data source, since no targeted elicitation of derivations is possible. Therefore, a collective representations analysis would require a rather large amount of such material, which might only

[15] The production of (longer) written primary texts in the research process usually fails due to the insufficient motivation of the persons studied (the writing of a written report etc. is far more time-consuming than, for example, answering questions in an oral interview). A more comprehensive text production can be achieved through stronger (financial) incentives; however, these bear the danger of a "substitute motivation" and thus an artefact production (if, in extreme cases, reports and narratives are freely invented in order to be paid for it).

be possible in an acceptable form by using archived audio files (e.g. of radio broadcasts).

4. Such a restriction does not apply to *"natural" written data material*, because the material basis seems almost infinite here. Newspapers, magazines, party programmes, leaflets, parliamentary minutes, specialist literature, non-fiction and advice literature, fiction, advertising texts and much more can form the source material for a reconstruction of collective representations. However, a considerable limitation arises from the fact that in this way it is possible that mainly publicly dominant representations are captured and made visible. However, the Internet in particular is making counter-discursive texts and interpretations increasingly accessible and usable for the analysis of collective representations.

A review of research papers with a collective representations analysis approach quickly reveals that the generation of verbal data material is the most widespread in the research process (cf. among others Bögelein and Vetter 2019 as well as Sect. 1.3). This can primarily be attributed to the aforementioned advantage of targeted elicitation of derivations. More recently, however, an increase in document analyses can be observed, which can be explained in particular by the growing discourse analytic research (cf. Keller 1997; Meuser 2010).

In order to generate data material from which derivations and collective representations can be reconstructed, qualitative interviews are used above all, and primarily guided interviews. Insofar as the form of the guided interview is further specified in this context, the Problem-Centered Interview (Alemann 2014; Bastian 2014; Bosancic 2014; Streckeisen et al. 2007; Weigelt 2010, among others) or the Discursive Interview (Alemann 2014; Bögelein 2016; Markova 2017; Sachweh 2010, among others) are usually used. In contrast, more narrative-generating interview forms such as the narrative and the biographical interview (e.g. Kunze 2011) are rarely found in a collective representations analysis approach.[16]

Group discussions are also a suitable instrument for recording collective representations. Here, the emerging group dynamics can be used to make collective

[16] "Expert interviews" (cf. Bogner et al. 2002; Gläser and Laudel 2009; Meuser and Nagel 1991, among others) are also used particularly frequently for research analysing collective representations. This is mainly due to the strong interest in reconstructing the interpretive knowledge of the respective expert type (Bogner and Menz 2001; Meuser 1992, among others). However, expert interviews are defined by the object of research. They do not represent a separate interview type and are mostly guided interviews.

representations visible. Group discussions seem to be particularly suitable for recording social structures of meaning, since collectively shared worlds of meaning and milieu specifics (i.e. also collective representations) become more apparent in such group discussions than in individual interviews, for example. Compared to interview techniques, group discussions also have the advantage that stronger interventions by the moderator of the discussion are not necessary, since corresponding incentives are provided by the participants of the group discussion themselves.

In the German-speaking world, the bulk of academic research with group discussions invokes Bohnsack's (1989, 1991, 2000) and Przyborski's (2004) conceptual considerations; Przyborski and Riegler (2010), which in turn derive primarily from Mannheim's sociology of knowledge.[17] However, this direction of the group discussion method is limited in its scope by the fact that they are only conducted with real groups or as far as conjunctive experiences (Mannheim 1980) can be assumed. In this way, typical group representations, which are shared by the members of the real group, are initially captured and reconstructed. The analysis of collective representations thus remains limited to such "conjunctive spaces of experience". Overarching collective representations as well as differences between collective representations below and across social groups cannot be ascertained in this way.[18] What is actually still to be determined, namely knowledge of the social distribution of collective representations, is therefore already assumed.

[17] Although this methodological school (of documentary interpretation) does not consider itself to be explicitly analytical of collective representations, it pursues similar objectives. Bohnsack (1992, 1997) uses the concept of orientation pattern, which has clear parallels to the concept of collective representations.

[18] Bohnsack, however, defines "conjunctive spaces of experience" relatively broadly and includes milieus and generations (1991, p. 115). This would argue against a restriction to real groups, which, however, is still the dominant group form of this research direction (cf. among others Bohnsack et al. 2007). Correspondingly, it could be argued that representations can also (or especially) be inferred with artificially composed groups (so e.g. Mensching 2010; Müller 2014). Here too, however, the problem arises that sufficient knowledge about the social distribution of collective representations must already exist in order to be able to compose the discussion groups. The difference between collective representations and those tied to "conjunctive spaces of experience", however, is primarily that collective representations are formed and validated not only through milieu-bound interactions, but also through public discourses.

1.2.2 The Reconstruction of Individual Derivations and Collective Representations

There is a widespread consensus that collective representations or derivations are to be reconstructed from the data material. However, there is no consensus on the question of which methods can best be used for these reconstructions. Lüders and Meuser already stated that "no specific procedure of data interpretation has been developed for the analysis of collective representations" (Lüders and Meuser 1997, p. 67). This has not changed and there is probably no reason why derivations and collective representations should not be reconstructed in different ways.

A look at the more recent research literature reveals a rather varied picture here as well, especially since several procedures are often used in one study. The hermeneutic procedures that are widespread in the German-speaking world are also most frequently found here. Thus, collective representations are reconstructed in particular with objective hermeneutics (Bastian 2014; Kunze 2011; Meuser 2010; Streckeisen et al. 2007; Wolde 2007), the documentary method (Müller 2014; Weigelt 2010), Mayring's content analysis (Alemann 2014; Scholz 2012), and grounded theory methodology (Alemann 2015; Wolde 2007). Seemingly, then, many, if not (almost) all, of the methods of social science hermeneutics (Hitzler and Honer 1997) are applicable to collective representations analysis. Despite all methodological and methodological differences, two essential basic strategies can be identified in these, which offer different starting points for collective representations analyses. These are comparing and sequence-analytical interpretation strategies.

Already at the end of the 1990s, Lüders and Meuser summed up that in a "manageable number of cases" the "sequence analysis has emerged as the suitable and so far hardly controversial procedure" (Lüders and Meuser 1997, p. 68). However, they do not explain which reasons speak for sequence-analytical interpretation procedures in the reconstruction of collective representations. In this respect, it can be assumed that the spread of sequence-analytical procedures is primarily due to the general success of objective hermeneutics (Oevermann), narrative analysis (Schütze) and the method of documentary interpretation (Bohnsack)[19] in German-language social research[20] – and not to the fact that collective representations can only be reconstructed by sequence analysis.

[19] Conversational analysis, which has been increasingly applied beyond its original context in recent times, has, in contrast, not yet had a major influence on collective representations analyses.

[20] For an overview and a comparative presentation of sequence-analytical methods, cf. esp. Kleemann et al. (2009).

A second reason is certainly the close "genealogical" connection between objective hermeneutics and the collective representations approach, which have a common "father" in Oevermann (including Oevermann 1973, 2001; Oevermann et al. 1979). The relationship between objective hermeneutics and collective representations analysis is, however, ambivalent to say the least (cf. Oevermann 2001; Plaß and Schetsche 2001). Above all, it is problematic that objective hermeneutics, due to the hypostasis of the concept of structure (cf. Reichertz 1988), has no theoretical or methodological means for the social localization of collective representations. A reconstruction of meaning structures carried out with objective hermeneutics or another sequence-analytical method will always only be a reconstruction – albeit a highly accurate and well substantiated one – of individual derivations. For a reconstruction of collective representations, the sequence-analytical procedures oriented towards case structures lack the instruments.

The alternative to sequence-analytical procedures are comparing analysis strategies. In these, the systematic (synoptic) comparison of text passages is the basic strategy of interpretation. In this process, parts of the text that are thematically similar but originate from different passages are contrasted (cf. Kelle and Kluge 1999, p. 56 ff.). In this way, these methods differ fundamentally from the aforementioned sequence-analytical ones, which understand texts as units of meaning and attempt to grasp meaning through interpretative tracing of the production process of a text. Comparative analysis strategies, on the other hand, aim at socially shared orientational knowledge. To this end, they use interpretive procedures that decidedly provide for such comparisons, such as the grounded theory methodology developed by Strauss and Corbin (Strauss 1991; Strauss and Corbin 1990; cf. also Charmaz 2014; Dey 1999; Strübing 2008), or at least allow for them, such as, in particular, content analysis according to Mayring (1983). Furthermore, comparative methods also allow for the exploration of a field of collective representations and are often explicitly aimed at (cf. Kelle and Kluge 1999, p. 75 ff.; Sect. 4.2). Finally, the use of software suitable for qualitative text analyses is typical and conditioned by the need to maintain an overview and control over seemingly "overflowing" quantities of text.

The fact that sequence analyses are nevertheless often regarded as having no alternative for the reconstruction of collective representations may also be due to the fact that comparing (or non-sequentialist) forms of interpretation are perceived as both "fuzzy" and methodologically "thin".[21] With the increasing reception and ap-

---

[21] For example, the attitude of many qualitative social researchers towards content analysis according to Mayring (1983), but also towards grounded theory methodology (Strauss and Corbin 1990) is often at least ambivalent. In some textbooks these methods are not presented at all or – measured against the requirements of sequence-analytical procedures – as deficient methods.

plication of grounded theory methodology (Strauss 1991; Strauss and Corbin 1990), which among other things attempts to methodically implement the "constant comparative method" of "old" grounded theory (Glaser and Strauss 1967), but also since Kelle and Kluge's (1999) presentation of contrasting and typologizing evaluation strategies, this perception should have changed. Critically, however, it remains to be noted that clear and methodologically as well as methodologically convincing accounts of comparing approaches are hard to find. (As justified in Chap. 2 and explained in more detail in Chap. 4, comparing is the central interpretative strategy of an analysis of collective representations with discursive interviews).

1.3 Analysis: Of Collective Representations: Research Examples

In order to give a better impression of the ways in which analyses of collective representations are carried out and the results obtained, this section will present some examples of research work. The selection focuses on studies that analyse collective representations (soziale Deutungsmuster) in a narrower sense. On the one hand, this excludes research that works with similar or competing concepts. The same applies to studies that are interested in collective representations, but for which their reconstruction is not the central concern. This applies in particular to studies with historical-genetic and discourse-analytical research interests, as well as to those that use standardized methods to investigate the social diffusion of representations.

Within this conceptual framework, the selection of the studies presented is then rather random and subjective, but attempts to take into account different methodological approaches and research fields. Therefore, it does not intend to be representative in any form, but only to provide a first insight into the practice of the research field of collective representations analysis. The presentation of the collective representations analysis studies takes place in short profiles or shorthand notes that follow a uniform pattern:

Thus, the research question or the research topic is first briefly introduced. If it emerges from the self-representations, it should also become clear why an analysis of (collective) representations is regarded as the right approach to the respective section of social life worlds of interest and how this is justified methodologically. In the second and third step, the procedure for data collection (From which material are collective representations reconstructed? How is material containing collective representations obtained?) and in data analysis (How are collective representations reconstructed from the previously obtained material?) will be explained. Particular emphasis is placed on the question of how the several analytical steps are justified in each case. The presentation of the research methods used is followed by a brief summary of central research results. In a final step, a brief theoretical and methodological classification of the studies analysing collective representations will be attempted.

A critique of collective representations analytic studies that goes beyond the narrower methodological and methodological context is not intended and should not be read into the shorthand notes. On the contrary, all the research presented here in chronological order impressively underlines, in my opinion, that the collective representations analytic perspective can produce high-quality and groundbreaking research results – and this even when methodological implementations are unclear or only very vaguely justified.

Brenke and Peter (1985): Arbeitslosigkeit im Meinungsbild der Bevölkerung [Unemployment in public opinion]. In: Michael von Klipstein; Burkhard Strümpel (Eds.): Gewandelte Werte, erstarrte Strukturen. Wie Bürger Wirtschaft und Arbeit erleben [Changed values, frozen structures. How citizens experience economy and work]. Bonn: Verlag Neue Gesellschaft. pp. 87–127	
Research topic and question	The study by K. Brenke and M. Peter deals with the (collective) representations (often just called "images") of unemployment and the unemployed. It examines which (collective) representations the perception of unemployment and the unemployed can be identified. The period under investigation is the early 1980s; it was conducted in West Germany (FRG). The authors do not refer to any literature analysing (collective) representations, but can best be attributed to the tradition of "workers' consciousness research"
Data collection	In addition to the qualitative part, the study also has a quantitative part (secondary analyses of survey data). With regard to the qualitative methodological procedure for recording the "unemployment images in the mirror of the qualitative interviews" (105 ff.), it is only mentioned that qualitative interviews were conducted (88), in which "the interviewees were able to present their views in their own words" (105). Further details on interview form or more concrete aspects (e.g. on a guideline) are missing. There is also a lack of any information on the number of interviews or the sampling procedure
Data analysis	Nor do Brenke and Peter provide any information on how the "qualitative interviews" were evaluated and how (collective) representations are reconstructed from the material.

(continued)

(continued)

Brenke and Peter (1985): Arbeitslosigkeit im Meinungsbild der Bevölkerung [Unemployment in public opinion]. In: Michael von Klipstein; Burkhard Strümpel (Eds.): Gewandelte Werte, erstarrte Strukturen. Wie Bürger Wirtschaft und Arbeit erleben [Changed values, frozen structures. How citizens experience economy and work]. Bonn: Verlag Neue Gesellschaft. pp. 87–127	
Central results	As a result, Brenke and Peter first note "a clear lack of orientation and uncertainty" and thus also that many respondents do not have or do not communicate clear (collective) representations with regard to both the causes and measures against unemployment (105 f.) In addition, the authors identify four groups of collective representations on unemployment: (1) the first comprises collective representations that see unemployment as a "consequence of undesirable economic developments". Here Brenke and Peter distinguish between three subtypes, which they call "reforms critical of capitalism" (including unemployment due to technical rationalisation), "market-oriented trade union criticism" (including wage levels as a cause of unemployment) and "imported unemployment" (unemployment as a result of wage competition and high commodity prices). (2) in the second type, unemployment is seen as a "consequence of the system", either political or economic. (3) in the third type of interpretation, "unemployment and economic crisis are seen as a consequence of wrong values" (e.g. selfishness/lack of solidarity; entitlement mentality). (4) finally, the fourth group comprises victimising (collective) representations which either see the cause of unemployment in the unemployed themselves or attribute it to job competition from migrants ("foreigners") (120 ff.)

(continued)

(continued)

Brenke and Peter (1985): Arbeitslosigkeit im Meinungsbild der Bevölkerung [Unemployment in public opinion]. In: Michael von Klipstein; Burkhard Strümpel (Eds.): Gewandelte Werte, erstarrte Strukturen. Wie Bürger Wirtschaft und Arbeit erleben [Changed values, frozen structures. How citizens experience economy and work]. Bonn: Verlag Neue Gesellschaft. pp. 87–127	
Methodological classification	The presented collective representations of unemployment are not very clear (types 2 and 3) and very heterogeneous (type 1 and type 4). As a result, the typology is not very clear-cut (e.g. types 1 and 2). Similarly, no systematic distinction is made between unemployment as a phenomenon, the effects of unemployment, the causes of unemployment and the perception of the unemployed Critically, however, it should be noted above all that methodological information is hardly available. Already this, but also many parts of the presentation itself suggest a rather low degree of methodological and methodological reflection. Therefore, the (collective) representations presented are more likely to be descriptions and summaries of interviewees' statements than interpretative reconstructions in the narrower sense At least in comparison to today's demands for reflection and documentation of the methodological procedure, the study by Brenke and Peter must be assessed as methodologically and methodologically "unsophisticated". Nevertheless, or perhaps precisely because of this, it also shows that the collective representations approach offers an interesting research perspective. Even in retrospect and with a distance of more than 30 years, the results of Brenke and Peter give an extremely illustrious impression of the perception of unemployment in the 1980s

Behnke, Cornelia; Meuser, Michael (1997): Zwischen aufgeklärter Doppelmoral und partnerschaftlicher Orientierung. Frauenbilder junger Männer [Between enlightened double standard and partnership orientation. Images of women by young men]. Zeitschrift für Sexualforschung 10: 1–18 Meuser (2010): Geschlecht und Männlichkeit. Soziologische Theorie und kulturelle Deutungsmuster [Gender and Masculinity. Sociological theory and cultural (collective) representations]. 3rd ed. Wiesbaden: VS-Verlag	
Research topic and question	The study presented in various publications was conducted in 1995 and, against the background of changing gender relations and discourses, dealt with young men's images of women, a hitherto neglected topic. The image of women held by young men is seen as an important indicator of their participation in processes of social change with regard to gender issues, partly because a "male-constructed" image of women is fundamental to the existing gender order
Data collection	The data collection was carried out with group discussions in the sense of Bohnsack (1991), which is justified by the fact that not individual attitudes but collective meaning are in the foreground of the interest of knowledge A total of 30 real groups (socially existing groups) with participants aged between 20 and 60 years were studied, which were located in different milieus and contexts (including skilled workers, gentlemen's club, students, "men's group"). The sampling was carried out as "theoretical sampling", so that new groups were still included in the study during the data analysis according to the principle of minimum and maximum contrast In the group discussions, the aim was for the discourse to be as "self-running" as possible, so that the participants could follow their relevance structures and focal points as unhindered as possible. The researchers' control of the discussion situation was therefore minimized and essentially limited to an initial stimulus ("what does it mean or what does it mean to you to be a man?")
Data analysis	The evaluation of the group discussions was oriented towards the "documentary method of interpretation" (Bohnsack 1991), which was originally developed specifically for the analysis of group discussions and therefore seems particularly suitable for the reconstruction of collectively shared (collective) representations of masculinity. However, the "comparative" aspect of this method, which is sequential at its core (a group-level comparison), is more accentuated. (The exact, step-by-step procedure or intermediate results are not presented)

(continued)

(continued)

Behnke, Cornelia; Meuser, Michael (1997): Zwischen aufgeklärter Doppelmoral und partnerschaftlicher Orientierung. Frauenbilder junger Männer [Between enlightened double standard and partnership orientation. Images of women by young men]. Zeitschrift für Sexualforschung 10: 1–18 Meuser (2010): Geschlecht und Männlichkeit. Soziologische Theorie und kulturelle Deutungsmuster [Gender and Masculinity. Sociological theory and cultural (collective) representations]. 3rd ed. Wiesbaden: VS-Verlag	
Central results	As a result, the authors identify on the one hand "habitual insecurities" among some groups of men, but on the other hand above all clear differences between the groups studied, ranging from "habitual security" oriented to old role models to "precarious security" to "institutionalized permanent reflection" and a "search for authenticity" Corresponding to these overarching basic patterns are other, sometimes contradictory interpretations and constructions of reality, which are reconstructed and presented in detail. Their shared background is the open or implicit confrontation with feminist social criticism and the demands for change derived from it Finally, Behnke and Meuser attribute differences between the groups (discussions) to milieu and socio-spatial contexts
Methodological classification	With regard to the research method used, but also with explicit reference to the different sociological theoretical directions and in particular to the sociology of knowledge of K. Mannheim, the own position is described as a sociological perspective of knowledge. Only by adopting such a perspective is it possible to grasp collective (as opposed to individual) contexts of meaning There is no explicit reference to a specific concept of collective representations; the concept of (collective) representations is rather used in a general sociological sense and is often replaced by terms that are obviously perceived as synonymous. However, due to the clearly reconstructive methodological approach and the self-positioning of the researchers, this study can clearly be attributed to the sociological analysis of collective representations

Rüling, Anneli (2007): Jenseits der Traditionalisierungsfallen. Wie Eltern sich Familien- und Erwerbsarbeit teilen. [Beyond the traditionalisation trap. How parents share family and employment]. Frankfurt/M.: Campus	
Research topic and question	A. Rüling's study deals with the strategies for reconciling family and work of (heterosexual) parental couples. The question is asked when a division of labour between parents is lived equally, which (collective) representations and strategies of action are effective and which structural resistances the couples encounter. The non-realization of equality (also) among egalitarian couples is a central research interest. It is assumed that the social framework favours more traditional settings ("traditionalisation traps"), which is why egalitarian couples have to develop specific strategies of action To answer these questions, parental couples were studied who implement an egalitarian division of labour at least in part. The reconstruction of the (collective) representations is intended to close a gap between the already well-researched area of the welfare state framework and the individual processes of action
Data collection	The study is based on data collected between 2001 and 2003. A total of 25 young parent couples were interviewed who attempt to implement an egalitarian (couple) division of labour. The procedure for selecting the parental couples can be understood as "selective sampling". Obtaining the interview data consisted of two parts: On the one hand, guided individual interviews were conducted with both parents, on the other hand, couple interviews The guided interviews were based on the Problem-Centered Interview (Witzel 1985) and the Discursive Interview (Ullrich 1999). The individual interviews focused primarily on the work situation of the interviewees as well as their daily routines and their private division of labour. In the couple interviews, the focus was on a joint assessment of the couple and their future life plans. In addition, the view of the family policy framework conditions was important here, which were also to be discussed, weighted and assessed on the basis of exemplary models
Data analysis	Already in the first evaluation step, couple types were formed on the basis of synoptic case descriptions. The second evaluation step dealt with the partnership negotiation processes and the handling of socio-political framework conditions. For this purpose, the coding techniques of the Grounded Theory Methodology (Strauss and Corbin 1990) were used. Finally, on the basis of the first evaluation steps, extreme cases of "traditionalisation traps" were identified and made the subject of in-depth case analyses in order to examine the different negotiation processes of the couples in greater depth and to clarify existing contradictions

(*continued*)

(continued)

Rüling, Anneli (2007): Jenseits der Traditionalisierungsfallen. Wie Eltern sich Familien- und Erwerbsarbeit teilen. [Beyond the traditionalisation trap. How parents share family and employment]. Frankfurt/M.: Campus	
Central results	Three traditionalisation traps are distinguished: (1) an impending risk of poverty due to the mother's return to work, (2) excessive demands due to the coordination of the professional development of both parents, and (3) gender-specific interpretations of childcare and housework. These structural settings correspond to specific (collective) representations and strategies of action. Due to structural obstacles (above all the socio-political privileging of the traditional division of labour), but also as a result of existing (collective) representations, egalitarian strategies of action are always unstable and must be constantly renewed and renegotiated. They are therefore always "exhausting" and associated with the danger of a relapse into traditional settings and couple patterns
Methodological classification	Methodologically, Rüling locates her study in the tradition of the interpretative paradigm, but avoids methodological and methodological specifications that go beyond this. Due to the strong accentuation of structural aspects, the orientation towards the structuring theory of Giddens (1997), the inclusion of macro-factors (social policy) as well as the theory-guided sampling, the procedure can, however, only partly be assessed as interpretative. This corresponds to the fact that the "found" (collective) representations are not systematically processed or contrasted at any point. It therefore remains unclear which (collective) representations were reconstructed and how they relate to the "traditionalization traps". It also remains largely unclear how the reconstruction of the (collective) representations was carried out methodically

Opielka, Michael; Müller, Matthias; Bendixen, Tim; Kreft, Jesco (2009): Grundeinkommen und Werteorientierungen. Eine empirische Analyse. [Basic income and value orientations. An empirical analysis]. Wiesbaden: VS-Verlag	
Research topic and question	The study by M. Opielka and colleagues examines the connection between the concept of basic income and value orientations. The aim is to clarify which values lead to support for a basic income (which does not yet exist as a social policy instrument) and which lead to rejection A central starting point is the observation that approval of a basic income cannot be read off from classical political attitudes. The reasons for the often reflexively negative judgements must therefore be sought at a "deeper" level than that of political attitudes, which, according to the authors, can be understood as (collective) representations. Assuming that future attitudes in the population are (pre-)shaped by the current assessments of the "elites", Opielka and colleagues attempt to record the (collective) representations of different social elites (business, politics, social work) that are relevant for the assessment of a basic income
Data collection	The research project comprises two forms of data collection, namely expert interviews and group discussions. The expert interviews "were intended to gain an insight into the interpretations of economic and political ethics in relation to the idea of a basic income among people who will shape the opinion picture in Germany in the near future" (43). To this end, interviews were conducted with persons at senior management level in the field of social and labour market policy. A total of 13 interviews with experts were realized, which lasted on average a little more than one hour. These expert interviews were "narratively designed"; in addition, a guide was used in the later course of the interview, which also contained confrontational questions In the second part, three group discussions (focus groups) were conducted with artificial groups assembled for the research. One group discussion each was conducted with (elite) representatives from the fields of business (8 participants), politics (8) and social work (6). In the focus groups, opposing actors were deliberately chosen in order to specifically work out "structural ambivalences of collective representations"

(*continued*)

(continued)

Opielka, Michael; Müller, Matthias; Bendixen, Tim; Kreft, Jesco (2009): Grundeinkommen und Werteorientierungen. Eine empirische Analyse. [Basic income and value orientations. An empirical analysis]. Wiesbaden: VS-Verlag	
Data analysis	Unfortunately, the way in which the data material was analysed is described only very vaguely and has to be "reconstructed" in part from the results. It seems clear that the evaluation, at least of the expert interviews, was essentially carried out using the Grounded Theory Methodology (Strauss 1991). Using qualitative analysis software, the material was coded and interpreted on the basis of these codes. For the evaluation methodology of the group discussions, it can be assumed that the documentary method (Bohnsack 1989) was also used, at least as a supplement
Central results	The central finding of the study by Opielka and colleagues is that different value concepts and perceptions (including ideas of justice and the concept of man) play a central role in the assessment of basic income concepts. The (collective) representations used to justify the rejection or advocacy of a basic income also differ significantly in some cases With regard to the three group discussions, there were also expected differences in the general assessment of a basic income: While the (elite) representatives of business consistently and categorically reject it, the "politicians" have a positive attitude towards it, while the representatives of social work show a rather ambivalent attitude towards a basic income
Methodological classification	The study by Oplielka and colleagues cannot be assigned to any methodological direction. The concept of (collective) representations is also used in a rather eclectic form, without addressing the contradictions that arise when referring to different methodological approaches. The (collective) representations described are correspondingly blurred. A systematizing comparison or a typifying condensation is not carried out. The methodological procedure is altogether only very briefly described, but already shows clear deficits. These relate in particular to the composition of the discussion groups, which could certainly be described as tendentious

Sachweh (2010): Deutungsmuster sozialer Ungleichheit. Wahrnehmung und Legitimation gesellschaftlicher Privilegierung und Benachteiligung [Collective representations and social inequality. Perception and legitimation of social privilege and disadvantage]. Frankfurt/M.: Campus	
Research topic and question	The research background of P. Sachweh's study is the increase in poverty and income inequality in Germany since the 1990s. The research interest is to clarify how (growing) social inequality is perceived, interpreted and also legitimized by people in different social situations. The research is guided by the assumption that this can only be explained by more complex patterns of social interpretation and that these must be reconstructed from material obtained in interviews
Data collection	The selection of the interviewees was partly carried out as "theoretical sampling". A selection criterion set by the research question, on the other hand, was the social or class situation, whereby the interviewees were recruited equally from particularly privileged and disadvantaged situations. This was based on the assumption that in these population groups the need to legitimise social inequalities is particularly pronounced. At the same time, other characteristics were also purposefully varied (e.g. gender and occupational situation) Discursive interviews (Ullrich 1999) were used as the data collection instrument because they are conceptually oriented towards the reconstruction of socially shared (collective) representations. The guiding principle in the conception and design of the interviews was to prompt the interviewees to make statements and give reasons by means of an elaborated guideline and specific questioning techniques, because these often fall back on (collective) representations In addition, procedures known mainly from the problem-centered interview (Witzel 1985) were used (memory protocols about the interview situation; short questionnaires with socio-demographic data). A total of 20 interviews with a length of about 60–90 min were conducted

(continued)

(continued)

Sachweh (2010): Deutungsmuster sozialer Ungleichheit. Wahrnehmung und Legitimation gesellschaftlicher Privilegierung und Benachteiligung [Collective representations and social inequality. Perception and legitimation of social privilege and disadvantage]. Frankfurt/M.: Campus	
Data analysis	The data analysis was oriented towards the procedure intended for discursive interviews and aimed at reconstructing collective representations from the individual (collective) representations or derivations in a contrasting analysis strategy. The analysis of the text material was carried out by means of continuous coding and the grouping of the codes into categories, whereby procedures of the Grounded Theory Methodology (Strauss and Corbin 1990) and the contrasting analysis according to Kelle and Kluge (1999) were used. Accordingly, both "deductive" (theoretical) and "inductive" (empirical) codes were used, which formed the basis for the comparative reconstructions of collective representations. Among other things, paraphrasing and explicative interpretation techniques (according to Mayring 1983) were also used in the process
Central results	The study arrives at a large number of partial results, among other things, on the manifestations and causes of social inequality. For the central question of the legitimation of inequality(ies), "semantics of justice" in particular proved to be fundamental. In particular, notions of performance-based justice, which in turn can differ significantly, seem to be shared almost universally. However, Sachweh was also able to establish the same, albeit somewhat weakened, for ideas of needs-based justice. Other concepts of justice and equality as well as non-normative interpretations, on the other hand, proved to be more socially differentiated
Methodological classification	The study is based on a rather sociological (and culturalistic) understanding of collective representations. It is assumed that (collective) representations can be characterized by inconsistencies (i.e. do not have to be logical or consistent) In the field of inequality research, the work sees itself in contrast to Marxist, functionalist and structuralist approaches, which do not problematise the question of the perception of inequality, or only marginally. However, it sees itself less as an alternative to classical explanations of social inequality than as a broadening of perspectives that takes a look at the representations of social inequality on a subject-oriented level

Bögelein (2016): Deutungsmuster von Strafe. Eine strafsoziologische Untersuchung am Beispiel der Geldstrafe [Collective representations of punishment. A sociological study of punishment using the example of fines.] Wiesbaden: Springer VS	
Research topic and question	The "sociology of punishment" work of N. Bögelein stands in a series of studies that deal with perceptions and perceptions of punishment (punitiveness research). Bögelein's object of research is the perception and assessment of fines. More specifically, she examines how people who have been sentenced to a fine perceive and evaluate their punishment and, above all, how they legitimize it for themselves. For this purpose, she resorts to a collective representations-analytical research design. A guiding idea is that a reconstruction of the (collective) representations of those sentenced to fines also allows conclusions to be drawn about socially widespread and generally shared (collective) representations of punishment The importance of the topic is derived from different theories of punishment. The fine is described as particularly relevant in this respect, as it is by far the most frequent form of punishment as well as the least studied and publicly negotiated, and is also controversial with regard to the purpose of the punishment
Data collection	The study is based on data from a detention avoidance evaluation project collected between 2011 and 2013. The sampling aimed at capturing all expected combinations of characteristics. The sampling criteria were thus already determined at the beginning of the study on the basis of theoretical considerations ("selective sampling"). Access to the target group (recruitment) took place through so-called gatekeepers (i.e. via institutions such as correctional facilities) and through self-selection (including the distribution of "advertising flyers") In order to obtain the interpretative material, guided interviews were conducted, which were predominantly based on the concept of the discursive interview (Ullrich 1999), but also contained elements of the problem-centred interview (Witzel 1985). A total of 44 interviews were conducted and analysed with individuals who, at the time of the interview, were paying off a fine (6), serving it with community service (6) or serving a substitute custodial sentence (32). The interviews tended to be short (14 to 66 min), which Bögelein attributes to a low "willingness to discourse" (2016: 127) on the part of her respondents. Other peculiarities of the interviewee type and possible effects on the response behaviour are also discussed in detail

(continued)

(continued)

Bögelein (2016): Deutungsmuster von Strafe. Eine strafsoziologische Untersuchung am Beispiel der Geldstrafe [Collective representations of punishment. A sociological study of punishment using the example of fines.] Wiesbaden: Springer VS	
Data analysis	The data evaluation was carried out in a multi-stage process and in a combination of contrasting and sequence-analytical procedures. Sequence-analytical fine analyses of a part of the interviews (9), carried out in accordance with Objective Hermeneutics (Oevermann et al. 1979), formed the basis for a case-contrasting reconstruction of collective representations. These (collective) representations were then traced in the remaining interviews by means of a structuring content analysis (Mayring 1983). Finally, (quasi)correlations between (collective) representations and types of crime were sought
Central results	Bögelein reconstructs six (collective) representations that the interviewees sentenced to fines fall back on: She divides these into (collective) representations with and without a moral dimension. Without a moral dimension, the (collective) representations are "money instead of punishment" (monetary interpretation), "punishment as fate" (fatalistic interpretation) and "punishment as risk calculation" (rational interpretation). With a moral dimension are the (collective) representations "legitimacy of punishment" (legitimate punishment), "punishment as compensatory justice and reparation" (just punishment), and "punishment as injustice" (unjust punishment) The individual collective representations in turn have their own dimensions, but are identified in their entirety as socially shared collective representations. Particular attention is paid to the contradictions that occur especially in the (collective) representations with a moral dimension, whose dynamics are understood as further evidence of a location in social knowledge bases
Methodological classification	The study does not explicitly assign itself to any methodological direction and attempts to combine above all objective-hermeneutic ideas and a sociological understanding of (collective) representations. Above all, the data collection procedure is explained and justified in great detail. On the other hand, the combination of interpretation and analysis strategies that are often regarded as methodologically incompatible (such as objective hermeneutics and Mayring's content analysis) must be regarded as "courageous", the justifications for which then turn out to be somewhat (too) pragmatic

Theoretical and Methodological Basic Assumptions of the Discursive Interview

2

Before Chaps. 3 and 4 present the concrete procedure for data collection and the reconstruction of collective representations, this second chapter first deals with the methodological considerations underlying the Discursive Interview. These are primarily intended to clarify how the Discursive Interview attempts to implement the conceptual clarifications of the first chapter and the resulting requirements for an analysis of collective representations. In doing so, it will always be necessary to go into general methodological questions a little more and, above all, to highlight differences to other forms of qualitative interviews.

First, the basic methodological premises of the discursive interview are explained (Sect. 2.1). The second subsection deals with the interview situation and with interviewing as a specific form of communication and aims to illustrate how the special features of the interview interaction can be used to elicit derivations in the discursive interview (Sect. 2.2). Finally, the central point for interview forms is dealt with, what questions in research interviews achieve, what types of answers they lead to and what type of answer the Discursive Interview aims for (Sect. 2.3).

2.1 Three Methodological Premises

The discursive interview is understood as a methodological procedure that encompasses all of the stages of an analysis of collective representations explained in Sect. 1.2: the acquisition of material from which collective representations can be reconstructed, the reconstruction of individual derivations, the reconstruction of collective representations through a contrasting analysis, and the mapping of the field of collective representations of an object area or reference problem. It is

C. G. Ullrich, *The Discursive Interview*,
https://doi.org/10.1007/978-3-658-38477-7_2

therefore conceived as an *integrated research design*: All research phases (sampling, form of questioning, contrasting interpretation and typologizing) are coordinated with each other in the Discursive Interview and form a functional unit that is oriented towards the reconstruction of collective representations.

In purely conceptual terms, the discursive interview does focus on data collection. However, considerations about "finding" collective representations (or material containing derivations) only arise from the analysis strategy and this, in turn, from the general research objective. Only after it has been decided to reconstruct collective representations by means of a contrasting interpretation, therefore, does the question arise as to a suitable form of data collection for this purpose, which is seen here in a "discursive" interview.

In this chapter, the general methodological considerations and reasons for using the discursive interview as a way for reconstructing collective representations are first presented. They will be discussed in more detail in the following chapters and, above all, with regard to their practical-methodological application.

The Discursive Interview is based on three key methodological considerations or *premises*, namely.

- on a basic understanding of the sociology of knowledge,
- on the communicativity of collective representations and their communicative elicitability; and
- on the conviction that collective representations can best be reconstructed comparatively.

2.1.1 Collective Representations as a Sociological Concept of Knowledge, but a Non-constructivist One

The theoretical and methodological position underlying the Discursive Interview is a sociological one.[1] It is neither positivist (at least not in the sense of the widespread use of this term) nor constructivist, insofar as more than a social constructivism is meant by this.

[1] Here is neither the appropriate place nor the necessary space to go into more detail about what is called sociology of knowledge or sociology of knowledge. For this, we can only refer to the relevant introductions (esp. Berger and Luckmann 1990; Knoblauch 2005). However, it should become clear from the following remarks that the sociological tradition of knowledge is meant in the succession of Schütz, Mannheim, and Berger and Luckmann, and that this sociology of knowledge has clear parallels in symbolic interactionism and ethnomethodology (cf. also Keller 2012).

For the analysis of collective representations, this means first of all that collective representations are social constructs. They are relatively independent of action situations and reference problems and never necessarily result from them. This relative autonomy of interpretations of only apparently objective problems becomes particularly clear in historical or comparative cultural perspectives. The strong change in the social perception of social problems (e.g. homosexuality, smoking, sexual abuse) can be regarded as almost classic here.

Even more: the socially available representations help to determine what is defined or perceived as a reference problem in the first place. Whether a situation is perceived by actors, for example, as dangerous, embarrassing, beautiful or simply as everyday and normal, and whether a need for action (and what action) is then derived from this, depends to a large extent on which collective representations actors can fall back on .

Problems of action or situation definitions are thus co-constituted by the socially available representations. However, this does not go so far that the situations and reference problems to which collective representations refer exist *exclusively* because of these collective representations. Rather, the situations and reference problems also have an existence independent of the respective interpretations with which they are provided, only that they are grasped and perceived by these in a certain way.

Only in this way does it also make sense to speak of an "experience dependence" (with relative "experience resistance") of collective representations (Sect. 1.2). In the case of lack of success or changing environments, individual representations are constantly readapted and this is only conceivable if such "experience testing" is possible. For example, what is considered poverty in a society, i.e. whether and in what a phenomenon described as poverty exists, is largely determined by the dominant collective representations (cf. Box in 1.1). Nevertheless, discourses and interpretations of poverty always also refer to material substrates (e.g. a certain level of income and provision).

Finally, collective representations are not a property of individuals, but part of a collective body of knowledge. Therefore, actors cannot simply design "their" collective representations or "rationally" choose between existing collective representations: Which collective representations are situationally appropriate or adequate is usually already "socially decided" and "imposed" on the agent, because this is a prerequisite for the action-enabling effect of collective representations. Only rarely, on the other hand, especially in the case of their failure, are collective representations consciously reflected upon. For this reason, the users or bearers of collective representations are not what the Discursive Interview is interested in, but exclusively the collective representations themselves. The interviewees are therefore "cases" only in the sense that they are users of collective representations; they are treated as "informants" about (or as access points to) them .

2.1.2 Collective Representations Must Be Communicated and Used to Justify Actions

As explained in Sect. 1.2, collective representations are as indispensable for orientation in social situations as they are for the justification of actions and decisions. Therefore, one will encounter collective representations in almost all texts. However, this will be particularly the case when texts contain a high proportion of justifications, explanations and argumentations. The discursive interview always offers itself as a method when such collective representations-intensive texts are not available to a sufficient extent, when specific collective representations of concrete social groups or types of persons are to be reconstructed, or when a targeted and systematic (controlled) variance is sought.

As a form of data collection, the discursive interview makes use of the fundamental communication and validation function of collective representations by aiming to generate justifications for one's own actions and judgments. The core strategy of the discursive interview is to provide a set of instruments for generating answers (texts or text types) that are particularly suitable for the reconstruction of collective representations. As will be explained in detail in Chap. 3, this consists of various questioning techniques that prompt interviewees to explain their actions and to valuate situations and reference problems.

These questioning techniques and the interest in justifications can be irritating for some qualitative researchers, because they seem to violate basic principles of qualitative research (cf. Sect. 2.3 and Chap. 3). Two things should therefore be emphasised here:

1. The discursive interview only consistently accomplishes what tends to be neglected in many forms of interview, namely that it is a conversation between two people and that the statements of the interviewee are always (also) a joint product of both participants in the interaction.
2. This, the interactivity or *reactivity* of data collection, is not attempted to be minimized in the discursive interview (this is the aim of all "naturalistic" interview recommendations as well as standardized interview techniques), but is the *prerequisite for the capturing of collective representations* and individual derivations in the first place. In this respect, there is no problem of social desirability here: in the discursive interview, "socially desirable" answers will certainly

occur; but: these are an expression of what the interviewees regard as communicable representations. They are therefore not a problem, but a logical consequence of the understanding of collective representations as it was outlined here.

However, this also means that the Discursive Interview is not interested in the "true" motives or in "authentic" narrations of interviewees, but exclusively in which interpretations and justifications they consider legitimate and therefore socially communicable. Thus, an analysis of collective representations with discursive interviews does not aim at the level of subjective meaning, but at the discursive universe. Whether interviewees actually orientate or have orientated themselves on the derivations revealed in the interview is irrelevant for this, although it is quite probable.

2.1.3 Contrasting Interpretation as a Way to Reconstruct Collective Representations

As part of the collective stock of knowledge, the individual user can always only be partially aware of his/her collective representations. Thus, collective representations manifest themselves in almost every verbal utterance (or at least every justification) in the form of derivations. However, these *are* not collective representations, but individual adaptations or parts of collective representations. It is therefore one of the basic considerations of the Discursive Interview that *collective representations* – in contrast to individual derivations – can only be reconstructed by means of a contrasting interpretative procedure. *Social meaning is* therefore not *revealed* through individual case analyses, but only *through the reconstruction and clarification of meaning in its sociality.*[2]

This is why the (termino)logical distinction between collective representations and individual derivations is so important. If one wants to reconstruct *collective representations*, one cannot be satisfied with case-specific reconstructions of subjective meaning: The capturing and reconstruction of individual derivations is only

[2] This is not to be understood in the sense of social accumulation or distribution of collective representations. Nor it is claimed that individual case analyses do not allow conclusions to be drawn about latent or generative structures. However, a reconstruction of the specifically *social* character of these structures of meaning requires a contrast (cf. on this in detail: Sect. 4.2.1).

one, albeit significant, intermediate step towards the reconstruction of collective representations.

Contrasting analyses can take place on different levels and at different points in the analysis process. As will be explained in detail in Chap. 4, in the Discursive Interview contrasting analyses are provided on three levels: for the reconstruction of the individual derivations of a reference problem, for the reconstruction of the collective representations and for the mapping of the entire field of all collective representations of a reference problem. The performance of these contrasting analyses requires data material that is correspondingly extensive and rich in contrast. Therefore, it must be ensured both in the sampling (number and characteristics of the persons interviewed), but above all in the conduct of the interviews, that there are sufficient possibilities for comparison.

2.2 The Discursive Interview as a Social Situation

The "research interview is a special kind of human communication" (Kohli 1978, p. 1). And as with all other forms of communication, it can be assumed that the form of communication has an impact on the nature and content of the "texts" produced in the process. Thus, it is also undisputed that the interaction process in which interview data is gained has an influence on the type of data obtained. The social and personality characteristics of interviewees and interviewers as well as the entire setting of the interview (e.g. the interview location) are among the most important influencing factors.

The fundamental importance of the interview context has always been known and acknowledged in qualitative social research. It has been periodically emphasized or even "rediscovered" since then (cf. among others Briggs 1986; Cicourel 1974; Deppermann 2013; Douglas 1985; Hoffmann-Riem 1980; Holstein and Gubrium 1995; Hopf 1978; Kohli 1978; Kvale 2007; Lucius-Hoene and Deppermann 2004; Mishler 1986). What consequences this insight has for the handling of interview data, however, often remains unclear or quite contrary conclusions are drawn from it.

With regard to the influences of the interview situation, two basic strategies can be distinguished in social research, which can be called conventional and reflexive, respectively. The conventional forms assume a *problematic influence of the interview situation*: Accordingly, all aspects of the interview situation can have an undesirable and uncontrollable influence on the answers of the interviewees. Strategies for minimizing this influence are therefore dominant here. Depending on the methodological position and the concrete research interests, these can be very different and range from the standardization of the interview in survey research to methodological rules for conducting the interview (e.g. in the narrative interview) to the demand for intensive interviewer training.

A second conventional way of dealing with this, opposite in consequences, is to deny or ignore this influence. Here it is argued that – in contrast to standardised surveys – in qualitative interviews the interview context either has no influence on the answers or the "right" one. This is justified by the "natural" character of the survey situation, which is modelled on or simulates an everyday conversation, and the "ecological validity" (Cicourel 1982) of qualitative interviews that this ensures. Thanks to the ("natural") form of communication, which is equivalent to everyday communication, respondents would express themselves in exactly the same way as they do in other forms of conversation in the everyday world.

Reflexive strategies are mostly based on constructivist and conversation analytic ideas. In contrast to the conventional ones, they acknowledge the *principal interaction dependency of the data obtained in interviews*. In this perspective, interview answers are always considered to be co-constructed by the interview situation. From this it is concluded above all that the interview context must be taken into account when interpreting interview data. In this context, "the conscious perception and involvement of the researcher and the communication with the 'researched' is taken as a constitutive element of the process of knowledge" (von Kardorff 1995, p. 4). This sounds plausible and simple, especially if the researchers proceed in a (self-)reflective way, as required (Breuer 2010), but it hardly seems to correspond to common interview practice. For beyond proclamative demands and individual exemplifications (e.g. Deppermann 2013), the bulk of qualitative interview research shows rather little interest in revealing the processes in which the research data are gained.[3]

Here, a third way of dealing with the interview situation is to be made clear for the discursive interview. It will be shown that the situation-dependency of data collection with qualitative interviews does not have to be understood as a problem or disturbance and that the interview interaction is not only important for the interpretation of the interviewee's answers. Rather, the discursive interview aims at actively and productively using the interaction dynamics of the interview for data collection, i.e. already in the process of interviewing itself (Sect. 2.2.3).

---

[3] This is true even for the methodologically most demanding interview form, the Narrative Interview (Schütze 1976a, b, 1977, 1983). Thus Mey (2000, p. 142) criticizes the Narrative Interview or Schütze for neglecting "situational aspects to a large extent, since he assumes that the dynamics of the narrative process prevail over the conditions of the current communication situation (i.e. the interview)".

In order to clarify the methodological considerations for this, however, the situational issue must first be presented in some detail. Its serious if not dramatic consequences, become clear when one deals with a widespread implication of the situationality or reactivity of interview data, namely the radical methodological critique of interviews that rejects them as a means of data collection (Sect. 2.2.1). The proposals for dealing with the situativity and reactivity of interviews in the field of qualitative methods and the difficulties that arise in doing so are discussed in Sect. 2.2.2.

2.2.1 The Situativity of Data Collection and the Radical Methodological Criticism of (Qualitative) Interviews

The fact that meanings or the statements of interviewees are not simply actualized in an interview, but in a certain sense only interactively generated, has led to two fundamental methodological critiques of the use of qualitative interviews (cf. a. Hammersley 2013, p. 67 ff.; Hammersley and Gomm 2008; Murphy et al. 1998, p. 120 ff.).

The older, comparatively moderate criticism, based primarily on *conversation analytic* considerations, makes interviews, like other reactive methods, appear to be a second-choice option (cf. inter alia Ten Have 2004, p. 56 ff.). Accordingly, interviews would inevitably be co-constructed by the interviewers simply through the need to ask questions. The information of the interviewees is therefore not "revealed" in the interview, but only "produced" in the interaction between interviewer and interviewee. Therefore, protocols of natural interactions (such as recorded conversations) are always preferable to interview transcripts as data material.[4]

This form of criticism clearly relativizes the possible use of interviews; however, at least in principle it accepts them as a method of data collection. The renunciation of interviews on the one hand (conversation-analytical) and their use on the other (interpretative) can be justified with different research goals (here: social construction processes; there: social meaning).

[4] Ethnomethodological conversation analysis is concerned with the social construction performances in and of conversations (i.e. including interviews). It is interested in how social order is produced in interaction processes and therefore predominantly analyses such very basic processes and mechanisms (such as the organisation of turn-taking or so-called repair actions). In contrast, it is not interested in the concrete content (and meaning) of the conversations studied (cf. e.g. Ten Have 2007). Consequently, conversation analysts deal with interviews and group discussions, if at all, only as a specific type of communication (cf. among others Baker 2002; Wolff and Puchta 2007). In other words, they ask how interviews and how meanings are interactively produced in interviews (but not which ones).

This changes fundamentally with the constructivist radicalization of the critique of the interview. For this *radical epistemological*[5] *critique* ultimately asserts nothing other than the methodological untenability of interviews as a means of data collection and justifies this by arguing that meanings are always and *only* constructed in the interaction situation.[6] "Answers" in interviews are therefore exclusively and no more than a product of a specific interview interaction and have no (independent) meaning beyond that. Another problem is that the interviewer's questions assume a certain reference to reality, a specific view of the social world. The interviewers thus inevitably have an influence that pre-structures reality and thus distorts it (cf. Potter and Hepburn 2005, p. 11 ff.).

The radical critique is thus directed against the idea – which is probably "self-evident" in all interview forms – that the data obtained provide access to the ideas of the interviewee or to his/her life and social world: "It challenges the use of these data as a window into the minds of informants and/or as giving access to information about the social worlds in which informants live." (Hammersley and Gomm 2008, p. 89).

If this is correct, interviews (of whatever kind) could indeed no longer be used conscientiously as instruments of data collection – or at best for rather preparatory and explorative tasks until more acceptable forms of data collection (can) be employed. Representatives of this radical interview critique (e.g. Potter and Hepburn 2005, p. 19 ff.) therefore also call for interviews either to be analysed throughout and only as social interactions or (better) to dispense with interviews and instead resort to recordings of natural conversations.

However, it is easy to show that this radical criticism of the interview is based on questionable premises and reductions and must be rejected:

1. First of all, it assumes in an almost naïve-positivist way that (interview-) external events and experiences are to be captured directly in interviews (Mishler 2005). However, reflection on the interaction context in qualitative interview methods shows that such a crude understanding of interviews as instruments of simple "data mining" (Kvale 2007) is unjustified. It corresponds neither to the

[5] Other forms of radical but non-epistemological criticism of (qualitative) interviews – in particular feminist criticism (e.g. Oakley 1981) and different variants of research ethics criticism (cf. e.g. Ten Have 2004, p. 72 ff.) – cannot be discussed here.

[6] "The radical critique of interviews as a means of collecting data about external reality (…) is seen by its proponents as the logical outcome of the understanding that interviews are essentially contextually situated social interactions" (Murphy et al. 1998, p. 120).

advanced interview methodology nor to the interpretative practice of many qualitative studies that resort to interviews as the mean for data collection.[7]

2. The accusation that interview responses are triggered first and only by the interview is also unconvincing. Of course, interviews are reactive techniques and the view that non-reactive procedures (in which researchers do not influence the social reality they are researching through the process of research) are preferable to reactive ones is probably one of the most widespread ideas in qualitative social research. On the other hand, at least all verbal data (i.e., including, for example, journal articles, diary entries, and recorded conversations) are almost always reactive (and, moreover, often intentional[8]) in that they refer directly or indirectly to other utterances. Whoever therefore rejects interviews as instruments of data collection solely on the basis of their reactivity would consequently have to refrain from using all other types of verbal data – or else justify this rejection on the grounds of an *interview-specific* reactivity.[9]

Finally, the reactivity critique – similar to that of the theory dependency or reality-constructing quality of interview questions – is based on a peculiar realism that is surprising, at least for a constructivist-oriented interview critique. For reactivity is only a problem as long as one assumes something like "true values," which are then not captured or distorted by the reactive approach. The heaviest cannons of constructivist interview criticism are thus, at their core, aimed at almost all research methods. They are not a critique of qualitative interviews, but a radical epistemology, unfounded in its radicalism, which in the last resort denies the epistemological value of any empirical research.

[7] Particularly elaborate here is certainly the Narrative Interview (Schütze 1976a, b), which is based on assumptions of narrative theory.

[8] The fact that (interview) statements can not only be reactive, but also or even only intentional (or pro-active), is in my opinion one of the most underestimated aspects in conventional interview research. For even if an answer in an interview is initially reactive, this does not exclude the possibility that respondents also "act" with their answers and, for example, purposefully set topics (cf. a. Holstein and Gubrium 1995). Ten Have (2004, p. 66 ff.) also uses exemplary analyses to show that interview questions are not only oriented prospectively towards possible answers, but also always have a "past reference" to preceding elements of the conversation.

[9] In contrast to radical interview criticism, one can even see an advantage in the specific reactivity of interviews. For this form of reactivity can – unlike the reactivity of so-called "natural" verbal data – be precisely observed in the research process and in part probably also controlled.

The third mistake of radical interview criticism consists in a blatant misinterpretation of interactionist and sociological postulates. It is true that one of their basic axioms is that meanings are created and changed interactively (cf. inter alia Blumer 1973) – but of course not only and above all not always completely anew. Through every interactive reference to interpretations, these may indeed be modified, even if often only incrementally. However, it by no means follows from this that in every interaction (and therefore also in interviews) meanings are always constructed completely anew and independently of problem references and socially shared knowledge bases.[10]

2.2.2 Interviews as Social Interaction

Even if inappropriate in its radicalism, the radical epistemological critique of qualitative interviews nevertheless sharpens the view for hitherto unsolved problems of interview methodology and for a perhaps too comfortable, self-contented interview practice. In view of the possibilities that interviews open up, however, even most critics rarely call for a complete banishment of interview methods. The typical claim then is that *interviews* should be *understood* as *social interactions and interpreted as a form of shared meaning constitution*,[11] especially through a complementary conversation analytic analysis of the interview interaction (see, among others, Deppermann 2013; Holstein and Gubrium 2002, 2011; Mishler 1986;

[10] To emphasize this – actually a sociological triviality – once clearly: Singular interaction situations (thus also a qualitative research interview) are neither sufficient nor necessary for the genesis of social meaning (construction of meaning). They are *not sufficient* because meanings are invariably "generated" in interactions with reference to (assumed) shared knowledge (and that almost always means: changed, but not completely newly created) and because meanings are always (have to be) tested, validated and transmitted through a multitude of interactions in order to become effective as social meaning. On the other hand, meanings are *not necessarily* dependent on concrete interactions, because collective representations (acquired, for example, in the socialization process) and other knowledge stocks can also simply be activated in singular interactions without being modified in more than an incremental way.

[11] The decisiveness with which this is demanded is irritating (and perhaps due to independently developing methodological discourses): After all, the view that interviews are interactions and that statements in (research) interviews can therefore not be understood detached from the interview interaction and the entire survey context has "always" been one of the central insights of qualitative interview methodology (cf. classically: Kohli 1978), without a "postmodern" methodological critique being necessary for this.

Talmy 2011). This would at least ensure the overcoming of a naïve positivist understanding of interviewing that only seeks to minimize the influence of the interview situation or tries to increase the quality of the interview data by collecting ethnographic contextual information. The systematic analysis of interview interaction and the consistent treatment of interview data as interaction products would not only prevent an uncritical treatment of information obtained in interviews, but would also lead to a more substantive interpretive outcome. It is said to be a matter of linking the "how" of generating meaning (here: in the form of a qualitative interview), which is central to conversation analysis, with the classically interpretative interest in the "what" (reconstruction of meanings).

This "*joint construction paradigm*" is relatively uncontroversial in recent interview methodology because, on the one hand, it seems to take the criticism of the interview seriously, but on the other hand it seems to avoid the radical consequence of abandoning interviews for data collection. Thus, it should be undisputed that the "interviewer contributions (...) must be represented and co-analyzed in the transcript as both a factually controlling, enabling and restricting, meaning-creating framework for the interviewees' actions" (Deppermann 2013, p. 20).

Neither this general insight nor the demands derived from it are really new (cf. among others Cicourel 1982; Holstein and Gubrium 1995; Mishler 1986; Silverman 1973). Unfortunately, however, the same is true of the problem central here, namely that in research practice it is hardly clear how this claim can be met. For the representatives of this position do not get beyond general plausibilizations and illustrative examplifications (for instance in Alvesson 2011, Deppermann 2013, Hester and Francis 1994, Holstein and Gubrium 1995, pp. 38 ff., Mishler 1986, pp. 52 ff., Talmy 2011 and Ten Have 2004). However, it would have to be clarified how and under which conditions the "how" of the constitution of meaning in the interview affects the "what" of its contents and how it is to be included the interpretation accordingly.

The analysis of research interviews as social interactions (or the requirement of such an analysis) is done with at least two different objectives:

The first is a kind of *withdrawal strategy*. Here, the skepticism towards an interaction-independent world leads to the renunciation of the exploration of primary experiences and to the turn to "second-order constructions". This strategy is thus not fundamentally different in research practice from ethnomethodological and conversation analytic positions. At its core, it consists of upgrading the interview interaction to a significant research object (cf. Baker 2002; Roulston 2014).

However, *validation strategies* are far more common (cf. among others Alvesson 2011; Brinkmann 2007; Holstein and Gubrium 1995). For example, Alvesson (2011, p. 129) justifies the parallel analysis of "how and what" by arguing that matching respondents' utterances with scripts specific to interview interactions counteracts objections to the validity of these statements. For to the extent that statements in interviews cannot be traced back to such scripts, they are a "strong indicator of how an interviewee experiences the focused social world". Precise analyses of the interview interaction are thus supposed to secure the interpretation of content, which for Alvesson remains the central concern of qualitative interviews. This claim corresponds to classical interpretative ideas and can be understood as the fulfilment of old methodological promises.

For the "validation strategy", in addition to the open pragmatic questions, a methodological problem also arises. For as soon as one distances oneself from the radical constructivist position that meanings are generated situationally and concedes an interview-independent world of meaning, it is also necessary to clarify why at all and how the concrete form of interaction (here, then: an interview) should affect the experiences and interpretations of the interviewees – and why and in what sense this then constitutes a problem.

Doubts about this are certainly warranted. Interviewers do restrict respondents by setting topics and by their questioning behaviour and thereby control *which* experiences, collective representations and knowledge are activated by the respondents. However, these are of course still the experiences, collective representations and knowledge stocks of these interviewees and do not arise ex nihilo in the interview interaction. The fact that interviews control or enable the activation of experiences and interpretations in the first place does not yet mean that this *must have an* effect on the meaning expressed in the interview (and must therefore be taken into account in subsequent interpretation). What is more likely and also sufficiently documented here is that the *type of activation* (e.g. the form of the question) – and not its "*whether*" – affects the *form of* a response, but not its *content*. In short, the interview interaction does not affect respondents' interpretations or experiences, but whether and how they are activated. Thus, a joint production of meanings in the interview interaction is not the same as a joint production of individual experiences, collective representations, expert knowledge, and so on. These are usually "linguistically mediated", but without being absorbed into language. They can therefore be reconstructed via and through the specific form of linguization.

The requirement to always analyse interviews as interactions thus seems more intuitively plausible than methodologically compelling. For one thing, interpretative and constructivist approaches and research goals are not necessarily methodologically compatible. For another, it would not only have to be illustrated for individual cases, but also justified in principle that and how the analysis of the "how" productively affects that of the "what". Otherwise, it remains unclear whether the interpretation of interview data as results of social interaction is meaningful or even necessary.

Table 2.1 Different methodological consequences of "joint construction" in interviews

Strategy	Methodological assessment of qualitative interviews
Withdrawal	Analysis (only) of the "joint construction" in the interview De facto rejection of interviews for the collection of "extra-situational" data Interest is limited to interviews as a form of interaction
Validation	Analysis of the "joint construction" to validate/triangulate the interpretive reconstructions Methodological ambiguity (of the "joint construction"-reconstruction) Practical vagueness (interpretation/meaning reconstruction)
Activation	Use of "joint construction" or interactivity already for gaining data Targeted interactive (discursive) interviewing Limitation of qualitative interviews to appropriate questions

By and large, then, two inferences from the constitution of meaning in the interview can be distinguished (see Table 2.1): the withdrawal strategy and the validation position, which is dominant in interview methodology. A third methodological position and strategies derived from it, which deals with "joint construction" both consistently and constructively, consists in not only taking into account the joint constitution of meaning in the interview interaction already during the interview itself, but also in using them for collecting data (see Table 2.1, row 4). This *activation strategy* is fundamental to the Discursive Interview and will therefore be presented in some detail in the following section.

2.2.3 The Active Use of Interview Interaction in the Discursive Interview

A more far-reaching implication of the situation-dependency of data obtained in research interviews than that of including them interpretatively is to use the interactional dynamics in the interview specifically to gain data.

So far, such a path has rarely been taken, and when usually only in part. Nevertheless, there is a series of preliminary considerations and studies to be drawn upon. These include analyses and research on interviewing as an activity (Hermanns 2000) or drama (Holstein and Gubrium 1995), on specific role expectations in the interview (Bogner and Menz 2001) or on the effect of specific question forms (Richardson et al. 1979). Particularly close are reflections on the subjectivity and reflexivity of researchers, which attempt to make clear that researchers or their subjectivity, which cannot be "switched off", are not necessarily to be regarded as a disturbance or source of error, but can also be a productive resource in the research process (cf. among others Breuer 2010; Breuer et al. 2011). However, an explicit concern with the data collection situation is an exception[12]:

It has already been shown in many studies that used a more "offensive" interview approach (e.g. Bellah et al. 1985; Bourdieu et al. 2002; Hochschild 1981) that an interview approach that is sensitive to the interactive data generating process can consciously and purposefully use the interaction dynamics between interviewer and interviewee for gaining data and lead to empirically interesting results. It is certainly no coincidence that these mostly proceed unconventionally, i.e. without an advanced methodological approach and without being attributable to specific interview methodologies.

A conscious approach to the shared constitution of meaning is the core idea of Holstein and Gubrium's (1995) "active interview", which they define not as an interview variant but a fundamental conception of qualitative interviewing. Central to this is the notion of an active role for both interviewers and respondents: "All interviews are interpretively active, implicating meaning-making practices in the part of both interviewers and respondents." (1995, p. 4). Just as respondents are not limited to a passive role, interviewers are not neutral instruments, but actively involved in data collection: "The active interviewer does far more than dispassionate questioning; he or she activates narrative production." (1995, p. 39). The influence of both interaction partners should therefore be taken into account in both the acquisition and the analysis of interview data (1995, p. 4).

This emphasis on an active interviewer role is groundbreaking. Holstein and Gubrium, however, remain mostly at the programmatic level and largely limit themselves to exemplary illustrations of techniques of active interviewing (1995, p. 38 ff.). They include "active listening", the use of background information, or the effect of preliminary information and opening questions on the respondent's answering behaviour. All these factors certainly point to an active interviewer role. However, they have been discussed in the interview literature for a long time and in this respect do not represent innovations (cf. esp. Briggs 1986; Douglas 1985;

[12] With group discussions, however, there is a survey method that explicitly uses interaction dynamics to obtain verbal data and differs from interviews precisely because of this. In particular, group discussions in the tradition of Mangold (1973), Bohnsack ((1991, and Bohnsack et al. 2007) are attempts to allow interaction dynamics to unfold in specificcontextual arrangements, which are intended to enable access to otherwise inaccessible areas of social reality. The aim is to make the group discussion as "self-running" as possible, with only a few interventions by the moderators. The interactive situation is thus used indirectly (passively) by relying on interaction and communication mechanisms that guarantee text production that is as independent of moderation as possible. In this way, this form of using interactive mechanisms differs fundamentally from the "active" one in interviews.

Mishler 1986). Most importantly, they tend to be expressions of unavoidable activity. For the question of the conscious and purposeful design of an active interviewer role, on the other hand, they offer hardly any clues.

Now, the discursive interview is understood as an attempt to derive specific techniques of targeted data gaining from assumptions about the mechanisms of interview interaction. The active role of the interviewer, the conscious shaping of the interview interaction (and not in the sense of a maximum "withdrawal" of the interviewer's influence) is derived from the specific research interest for which it was developed, i.e. the capturing and reconstruction of collective representations.

Based on the methodological background assumption explained in Sect. 2.1 that collective representations are primarily recognisable in justifications, the discursive interview aims to "elicit" individual derivations through questions and other requests for explanations. Interviews basically offer favourable conditions for this, because questions create a corresponding pressure of expectation (Ten Have 2004, p. 66 ff.).

The discursive interview goes beyond other forms of qualitative interviews in that it provides for explicit questioning techniques and appropriate interviewing for the elicitation of individual derivations. In this sense, the interviewers should have an activating effect and participate very consciously and necessarily in the process of eliciting derivations through different types of speech prompts. In addition, they can modify their role during the interview and, for example, switch from a more restrained to a more discursive interview conduct.

While Holstein and Gubrium (1995) used the "active interview" to illustrate the inevitability, but at the same time also the usability of active interviewing, the discursive interview thus offers methodological reasons for active forms of interviewing in relation to a specific epistemological goal. As a method based on the sociology of knowledge, the Discursive Interview can thereby dispense with postulates that are central to many directions of qualitative social research. As explained in Sect. 2.1, the discursive interview or the analysis of collective representations does not require or even aim to obtain authentic information about interviewees and their social world. For in this perspective it is not of interest what the interviewees "really" mean, think or feel, but "only" what they (can) communicate as opinions, thoughts and feelings.[13]

[13] With some justification, one can claim that, in the final analysis, actors can only subjectively mean, think and feel that for which they have presentations or derivations. Whether emotions, for example, actually exist (or can exist) only "linguistically", is a question that extends far into assumptions about the human condition and cannot, of course, be decided here. What is important for the position represented here, however, is that emotions and interpretations can only be communicated and attain social relevance to the extent that they are linguistically (ergo: social) constituted.

As long as the interviewers themselves do not "suggest" collective representations by offering them rashly as options for interpretation, it can therefore be assumed that the derivations expressed by the interviewees are authentic in the sense that they believe they can use them to explain their actions and definitions of the situation to third parties (in this case the interviewer(s)). Therefore, the interview does not result in a "joint constitution of meaning", but in an interactive activation (jointly triggered by interviewer and interviewee) of individual derivations of collective representations. Not a common genesis of meaning, but the interactive elicitation of meaning (cf. also Cicourel 2005) is the goal and the result of discursive interviews. Although this does not release us from the necessity of capturing this interactional process interpretatively, it does relieve us of the suspicion of a purely interview-situational and correspondingly limited or even questionable genesis of meaning. The problem of interpretation thus shifts from the joint constitution of meaning to the question of whether collective representations were activated in an adequate way.

Further Prerequisites for a Methodological Foundation of Active Interviewing

The Discursive Interview derives active interviewing from the goal of eliciting individual derivations. A general and fundamental methodological treatment of the influence of the elicitation context on data genesis in interviews is still far away. At least four sets of questions would have to be clarified for this purpose.

1. First of all, it should be clarified how the *interaction type "qualitative interview"* differs from others and in particular from those in which "questions" and "answers" as well as a corresponding distribution of roles are central (e.g. job interviews, examinations, interrogations). In order to clarify these questions, linguistic and communication-theoretical concepts can be used in particular. Possible starting points include conversation analytic work, interaction analyses in the tradition of Goffman and Schütz, or Gumperz's (1992) concept of contextualization cues.
2. Secondly, it would have to be clarified how the different *qualitative interview methods* differ with regard to their conceptions of interviewer and interviewee roles and the resulting consequences for the course of the interaction (including topic setting, conversation control, questions).
3. Thirdly, it would have to be analysed in more detail how different *interview settings* affect the interview interaction. The interview setting in-

(*continued*)

(continued)

cludes in particular the spatial conditions, characteristics of the interviewers and interviewees (and here in particular also differences between interviewers and interviewees, e.g. with regard to gender, age or social status) as well as special features of the interview situation (e.g. interruptions). Above all, empirical comparative analyses of different interview settings would provide information about the effects of situational contexts.[14]

4. A fourth level of investigation is that of specific elements of conversation, such as *questions* and other *prompts in* particular. Since these are of central importance for the Discursive Interview, their role in the process of data collection will be discussed in detail in the next section.

2.3 Questions and Answers in Qualitative Interviews

In the following, we will examine what effects questions and other conversational techniques in qualitative interviews have on the "answers" of the interviewees.[15] Behind this is the much more fundamental question of why we ask questions or inquire about something at all (in interviews or in other situations), that is, why (with what expectation) we think that by asking questions we can trigger one or more types of responses (more precisely: speech acts) that contain significant information for us.

As shown in Sect. 2.2, interviews are a form of reactive data gaining. Therefore, everything that is said in an interview by an interviewee is both part of and the result of a specific interaction or communication situation, i.e. it does not exist independently of

[14] Comparisons of classic face-to-face interviews with (qualitative) written surveys (Schiek 2014), online interviews (Früh 2000; Jones 2004), and other "low-context" survey situations can also be informative here (cf. Schiek and Ullrich 2016).

[15] The following remarks focus on areas that are central to qualitative research interviews. In addition, there are various other disciplinary approaches and interests in "questions" that can be significant for the effect of questions in research interviews. These include journalistic interviews (e.g. Haller 2001), therapeutic and counselling interviews (e.g. Rogers 1993; Wolff 1986), court proceedings and interrogation techniques, communication theories and models (e.g. Bühler 1999), the logic of questions (e.g. Belnap and Steel 1976; Ladanyi 1965), and the 'ethnography of communication' (e.g. Goody 1978; Hymes 1974). On the difference between (social science) research interviews and other forms of questioning from the perspective of survey research, cf. e.g.. Meulemann (1993).

this situation. All statements that interviewees make in the interview would – without this interview situation – either never be made or at least not exactly in this form.

Most, if not all, statements made by interviewees are first and foremost answers to questions posed by interviewers, responses to corresponding requests to speak. This is not to be understood as meaning that questions necessarily lead to certain forms of answers. Interviewees always have more or less leeway in how they respond to questions. Even more, the interpretative performances of the interviewees are a necessary prerequisite for the success of question-answer communication, as Goffman (1981b) in particular has made clear (cf. also McHoul 1987).

Nevertheless, it is maintained here that questions also have a situation- and recipient-independent influence on the nature of the answer. For even if respondents are free in their answers and may (and should) follow their own relevances, there is no doubt that the interviewer's questions pre-structure respondents' answers, albeit to varying degrees. Respondents are at least significantly less likely to comment on things that are not suggested by the interviewer's questions than they are to comment on topics that are directly or indirectly asked about.[16] Questions and their potential effects are thus crucial to what data are obtained in research processes that rely on interviews for data collection.

2.3.1 Types and Effects of Interview Questions

What questions are might not seem to need further explanation at first. After all, questions, asking questions as an activity and being asked questions are part of everyday experience. But of course it is not that simple. For one can easily convince oneself that we refer to quite different things as "question" and that questions often have great similarities with "non-questions". Therefore, it is first necessary to clarify what is meant here by questions.

With a view to how interview questions affect the answers of interviewees and what differences are possible, the following definitions seem to be useful:

1. First of all, in qualitative interviews there is an interest in verbal responses that are reactions to interview questions, and therefore in *speech prompts* (more precisely: non-binding[17] speech prompts). This excludes, among other things, rhetorical questions and questions that call for action (e.g., "Can you help me?").

[16] As discussed in Sect. 2.2, this feeds the general suspicion of radical interview criticism that interview statements are exclusively products of the interview situation and have no reality content beyond that.

[17] This term goes back to Hindelang (1994, p. 58 ff.), who distinguishes questions from other speech requests (e.g. commands).

2. For the interest in verbal responses, it is also important that the function of a speech prompt in terms of the effect it is intended to have is basically independent of its formal structure (see Table 2.2). Often, responses in qualitative interviews are triggered by other forms of (non-binding) speech prompts. These will be referred to and treated here as question-equivalent and *question-equivalent speech prompts*, respectively. Any differences between questions and question-equivalent speech prompts (usually: "requests") with respect to their response effect are not a conceptual but an empirical question that would have to be investigated first. What is decisive here, then, is whether it is a question in the functional sense (and not in the formal sense) in the sense of a speech prompt.

In addition, only "*genuine*", *information-generating questions* (and prompts) that are intended to elicit "informative answers" from the addressees are treated as questions and question-equivalent prompts in the following. These information-generating questions in the broader sense, which are thus used directly for data collection in research interviews, are the focus of the methodological considerations of the Discursive Interview.[18]

In the following, questions or "questions and other speech prompts" are thus non-binding speech prompts that aim at an informative answer, even if they would not be considered a question under formal aspects.

Table 2.2 Formal and functional questions and speech prompts

	Formal: questions	Formal: no questions
Functional: questions	**Information generating questions**	(Information-generating) **question-equivalent speech prompts**
Functional: no questions	Among other things, rhetorical questions and calls to action	No questions (e.g. statements)

[18] It is undisputed that other (formal) questions and speech prompts as well as the wider interview situation are also an important part of the overall interview interaction. This is especially true for the organisation of the interview communication (including motivating the interviewee, filtering, repairing). These interview techniques have already been studied relatively intensively, especially in conversation analytic and linguistic studies. They should be clearly distinguished from "information generating" questions.

On the Importance of Questions in Qualitative Interviews

In the methodological literature, the need for a thorough and methodological development of interview questions is repeatedly emphasized. The emphasis on the importance of questions is at odds with an often lacking methodological reflection. Thus, a look at teaching and handbook literature on qualitative interviewing (including Gläser and Laudel 2009; Gubrium and Holstein 2012; Helfferich 2009; King and Horrock 2010; Kruse 2014; Kvale and Brinkmann 2009; Reinders 2005; Seidman 1998) reveals the low status of the impact of questions in qualitative interview methodology. Questioning techniques and different forms of questions are mostly discussed here only in connection with practical issues (guideline construction, conducting an interview). The recommendations, which may well differ in detail, usually take the form of rules of thumb that are either not justified at all or based on practical research considerations and experience (cf. among others Helfferich 2009, p. 108; Kruse 2014, p. 215 ff.). Attempts to classify interview questions (e.g. Gläser and Laudel 2009, p. 130) appear correspondingly arbitrary due to the lack of methodological discussion and are based on heterogeneous, often incommensurable criteria (formal, conversational, functional).

In the German-language literature, Hopf (1978) analyses of a "guideline bureaucracy" (Leitfadenbürokratie) are the most important reference point for questions in qualitative interviews. In this context, Hopf deals with problems of interviewing that arise especially when using guidelines such as ignoring relevant points (1978, p. 103), a too "rapid change of topics" (1978, p. 105) or also dysfunctional role behaviour (1978, p. 110 f.). In this context, she also discusses problematic question formulations such as unclear and too long questions, overloaded questions, and suggestive formulations (1978, p. 108 ff.). Her considerations closely follow the criteria also known from survey research and the results of Richardson et al. (1979) and do not refer to the factual-contentual function of questions.

To Sensitize Interview Techniques to Questions and Question Effects

A stronger interest in the impact of questions can be expected above all from the Problem-Centered Interview, which not only provides for different questioning and stimulus techniques, but also openly advocates them (cf. among others Mey 2000; Witzel 1985; Witzel and Reiter 2012). And indeed, the considerations developed by Witzel for the Problem-centered interview on interview design and in particular on means and strategies of interviewing still represent the most important contribution to a systematic consideration of interview techniques and types of questioning. The different questioning techniques are thereby advised as strategies for deepening understanding, for motivating further explica-

(*continued*)

(continued)

tions and for avoiding or clarifying misunderstandings (Witzel and Reiter 2012, p. 75 ff.). These recommendations are presumably based on the empirical knowledge of the authors and users; in any case, no further methodological justification is provided as to why these effects actually occur. Thus, the reference to targeted response types or text types often remains vague, although the Problem-Centered Interview is avowedly interested in different contents – besides narratives at least still in "information" and validations.

More explicitly oriented towards text features and their analysis is the Narrative Interview (Schütze 1976a, 1987), which is based, among other things, on narrative theory (Labov 1972; Labov and Waletzky 1967). The central concern of the Narrative Interview is to elicit narrations through specific conversational techniques, among which the "narrative prompt" opening the interview is the central element. Questions and other stimuli are therefore discussed in terms of precisely this quality (How well do they elicit narrations? How well do the interviewer's statements support the interviewee's narration?) For narrative analysis it is then important to distinguish narrations from other text types (cf. Sect. 2.3.2). By its very nature, however, it is not interested in (or only delimits and does not further differentiate) other answer or text types –especially not in which questions and stimuli are necessary to produce text types other than narrations.

As illustrated in Sect. 2.2, recent discussion has been dominated by postmodernist and constructivist critiques of qualitative interviews (see Fontana 2002; Hammersley and Gomm 2008; Potter and Hepburn 2005). Among other things, this has led to a greater awareness of the "joint construction" of meanings by interviewers and interviewees. In this perspective, it would have been quite obvious to go beyond the emphasis on the importance of the interview situation to considerations of the effect of concrete elements of interaction – and especially that of questions. However, this is not the case. For this interview-critical discussion remains "in principle" or at the level of the general interview interaction. It has not led to a more intensive consideration of the effect of questions or different forms of questions.[19]

[19]Thus, at least on this point, the more recent methodological discussion falls behind the level already reached in the mid-1980s (cf. esp. Briggs 1986; Mishler 1986). In his linguistically influenced reflections on qualitative interviews, Briggs (1986) in particular had dealt with the communicative prerequisites and problems of interviewing. Among other things, he examined the significance of role expectations in the interview, of (question) contexts, of cultural differences in communication norms, and of the indexicality of verbal utterances for interview communication.

The fact that we know little about the effect of questions and other speech prompts (both in general and in qualitative interviews) is mainly due to the fact that their effects have hardly been empirically studied so far. For example, the only known empirical-experimental study of question effects in the context of qualitative interviewing dates back to the 1960s. It was conducted by Richardson and his colleagues (Richardson et al. 1965, 1979) and dealt in particular with the effect of leading questions, in which they distinguished between expectations and premises. An unexpected finding of this study was that leading questions were no more likely led to distorted responses than other questions (1979, p. 220).[20] One of their conclusions, therefore, is that leading questions are not problematic in principle, but only when they are uninformed and therefore imply something wrong.

Richardson et al. (1965), however, also omitted to analyze the leading questions they identified with regard to their wording structure. The only distinguishing criterion remains "informedness" (i.e. a content-related criterion). Richardson et al. (1965) do not provide any further analysis of whether or how leading questions have specific and possibly different effects due to their wording.

These and other research findings by Richardson et al. (1965) provide important indications for the questioning techniques of the discursive interview and in some cases also a basis of legitimation for maybe initially irritating forms of questioning (see Chap. 3). It is therefore all the more regrettable that such attempts to specifically investigate the effect of questions in research interviews have not yet been continued.[21]

On the whole, therefore, even the recent increase in interest in research interviews as a form of interaction has not led to a more intensive discussion of the types of questions and speech prompts that are used to elicit which types of re-

---

[20] Unfortunately, the study of Richardson et al. (1965) is not documented. Therefore, we do not know how distortions of responses were detected.

[21] Survey research has been concerned with the effect of questions (and the entire interview situation) on response behaviour longer and more intensively than qualitative interview methodology. The focus is on aspects such as question order, question and answer forms, social desirability, acquiescence and non-response (cf. among others Diekmann 1999, p. 371 ff.; Meulemann 1993). Which type of response is generated by which type of question, however, is outside the interest of survey research. This is also true for interactionist and conversation analytic analyses of standardized surveys (Fink 1995; Foddy 1993; Houtkoop-Steenstra 2000; Maynard et al. 2002). Primarily, this can be attributed to the fact that in standardized surveys no text types are generated and that the standardization of the answers already decides during the questionnaire construction which type of answer and which contents are possible.

sponses. This is already true for methodological considerations, but even more so for empirical studies on question effects.[22]

What Do Questions Do? Interaction Theoretical and Linguistic Analyses

However, questions (and answers) are not only important for research methodology, but are naturally also examined from completely different perspectives. Especially from conversation analytic, interactionist and linguistic analyses of questions and conversations, impulses for the design of questions in qualitative interviews can be expected.

For example, ethnomethodological conversation analysis is interested in the processes of social interaction, but not in interviews as instruments of data collection. Nevertheless, conversation analytic studies are also decidedly concerned with questions (e.g. Bergmann 1981) and even with questions in (standardized) research interviews (Heritage 2002; Schegloff 2002). However, they do so with a specific focus, namely that of conversational organization, and are therefore interested in such questions and mechanisms that are necessary for ordering communication (including conversation opening, turn-taking, and repair actions). Conversation analytic interest is thus always directed at the form of communication; questions and their effects come into view only to the extent that they function in the sense of communicative success. The effect of thematic questions, on which kind of answers (text types) they aim, lies outside the research interest of conversation analysis.

Goffman adopts a similar perspective in his interaction analyses (Goffman 1977, 1982, 1986), especially when conversations are their central subject (esp. Goffman 1981a). Thus, Goffman (1981b) subjects questions (or "statements") and answers to a precise examination, which deals in particular with conversation analysis and the concept of "adjacency pairs" and also includes speech act theory elements. He distinguishes "replies" (roughly: verbal answers to a question) as a sub-form of "responses" (roughly: reactions to requests to speak and act). Similar to the conversation analysts, Goffman is also primarily concerned with basal interactional processes. For this reason, but certainly also because his main focus is on the "responses", he omits to examine and differentiate the verbal answers ("replies") in more detail. In his analyses, Goffman therefore also does not deal with the types of response generated by questions.

[22] This low level of interest in the effect of questions may be due to the idea that possible problems can already be overcome by simple sensitisation (using "not to do" lists) and by training the interviewers. Also, that belonging to the same cultural group as the interviewees, general social skills or even just a social science degree already sufficiently qualifies one to conduct qualitative interviews seem to be widespread ideas still (cf. Hopf 2002, p. 357 ff.).

Questions are also an original object of linguistic analysis (cf. inter alia Conrad 1978; Hundsnurscher 1975; Walther 1975; Yang 2003), without, however, a uniform understanding about questions and especially not about different question types and classifications. Empirical linguistic studies on the use of questions and their effects are rare exceptions in this respect. The function of questions is usually derived from their formal structure (cf. among others Hindelang 1995; Vandeweghe 1977; Wagner 2001).

The same applies to speech act theoretical approaches (cf. among others Hindelang 1980, 1994; Klinke 1975; Searle 1971; Wunderlich 1976). Here, too, empirical studies on the effect of questions are the exception (but cf. e.g. Rost-Roth 2006). However, Sikora (2012) was at least able to show that the communicative functions of questions and different types of questions (such as insinuation, criticism, doubt) can in principle be reconstructed empirically. In doing so, he also makes clear that the same or similar questions can have different communicative functions depending on the context or call for different speech acts.

Overall, it is clear that in order to assess the impact of questions in research interviews, one cannot rely solely on semantic analyses and categorizations of questions or on conversation analytic and interactionist analyses of the general interview interaction. This is because linguistic analyses focus either on the formal structure of questions or on the analysis of the entire interview situation. There is very little interest in *empirical* research on question effects in these fields. But precisely this – empirical research on the effects of questions (in interviews and other conversational situations) – would be necessary in order to be able to assess their actual effects. In order to justify the forms of questions in the discursive interview, it is therefore necessary for the time being to resort to general considerations and theoretical assumptions (see Chap. 3).

2.3.2 Answers and Text Types

Like questions, answers (responses) are an inconsistently used category with many forms and needs for differentiation (cf. among others Goffman 1981b; Lang 1993; Waldenfels 1994; Wunderlich 1976). Answers can be treated as all verbal-informative utterances that are a reaction to a previously posed question. This stipulation excludes, for example, (pure) actions and bodily reactions as answers, even if they are in response to questions or other speech prompts.[23] The same also ap-

[23] In contrast, actions and physical reactions (especially facial expressions and gestures) can be important contextual information of verbal responses and should be taken into account accordingly, if documented.

plies to verbal utterances that cannot be considered a reaction to a question because, for example, they are made by the interviewee without being asked. Replies that are not answers in the sense of the question, but are to be understood as a reaction to it (e.g. in the case of misunderstood questions), on the other hand, should be included if they are of a verbal-informative nature.

In the literature on interview methodology, it is very common to see the goal of interviews in the acquisition of "information", without further explaining what is understood as information. In principle, this is certainly correct, if information is generally defined as "data to fill a gap in knowledge". However, this is of course an extremely vague definition, because what is considered to be a knowledge gap can vary considerably, and qualitative interviews – in contrast to interrogations or journalistic interviews, for example – usually go far beyond closing simple knowledge gaps.

However, an understanding of interviews as instruments for gaining information seems too simple – after all, it suggests that the contents of a answer is all to be found out with qualitative interviews, when in fact "only" data material is usually obtained, from which collective representations or other forms of meaning are (have to be) reconstructed in a complex analysis. It can therefore be assumed that qualitative interviews are usually not only intended to gain specific information, but that more complex knowledge formations are the central goal. In order to gain this knowledge, the interviewees in qualitative interviews are motivated to longer answers. These can be described and analysed by various characteristics. The most important feature, at least for the discursive interview, and presumably also the one most influenced by the type of questions or speech prompts, is the *text type*. In this sense, it can be said that specific answer text types are targeted with interview questions.[24] It is therefore assumed here that the "answer texts" obtained in qualitative interviews can be well and meaningfully distinguished on the basis of the text types they contain, which are stimulated with different questions and speech prompts.

---

[24] "Text type," however, is a feature used to describe texts that is as common as it is variously defined, and is therefore anything but unambiguous (cf. Adamzik 1995, 2008; Heinemann 2000, Heinemann and Heinemann 2002; Isenberg 1978, among others). Complicating matters further is the fact that similar text features are often referred to by different terms (e.g., text pattern). If the term "response or text type" is used for the characteristics of interviewee responses of interest here, and thus given a rather specific meaning, this is done primarily because this use of the term can be linked to a widespread practice in the field of qualitative social research. Thus, the analysis of text types is mainly known from narrative analysis (Kallmeyer and Schütze 1977; Schütze 1976a, 1987) and procedures based on it (e.g. Fischer-Rosenthal and Rosenthal 1997).

Based on methodological literature on interviews (esp. Schütze 1976a, b, 1977; Ullrich 1999; Witzel 1985), several such text types can be distinguished, in particular narrations, descriptions, justifications, reflections, information, explications and validations (see Table 2.3, column 1). In order to activate these, different types of questioning techniques are possible and have been suggested (see Table 2.3, columns 2 and 3), which create a corresponding pressure of expectation on the part of the respondents (column 4).[25]

Text Types and Text Type Elements

At this point, it is unfortunately unavoidable to introduce a further differentiation, namely that between text types (TT) and text type elements (TTE). The term text type is used here for larger text units (see Table 2.3). Most questions and prompts are aimed at this level and are intended to evoke narrations, explanations and other text types.

However, these text types are in turn composed of different text type elements. For example, a narration consists not only of narrative but also of many non-narrative parts such as descriptions or justifications (cf. Quasthoff 1979), and justifications can also contain narrative or descriptive parts (Antaki 1994; Kienpointner 1992). (Other text types such as explications and validations even have no "own" text type element defining them and consist of a typical combination of text type elements). Thus, not only may an answer given in an interview comprise several text types, but these text types will in turn often be composed of several text type elements.

Table 2.3 distinguishes, as a preliminary classification, seven text types (TT): Description, Narration, Justification, Information, Validation, Explication and Reflection. All text types usually contain several different text type elements (TTE). Other text types, which are probably less important for the extraction of derivations, are e.g. explanations/instructions or reports. The number of possible text type elements will be considerably larger, because in addition to text type-typical ones (descriptive texts, narrative texts, justification texts/arguments, "pieces of information") there is also a certain number of non-specific text type elements (including assertions, summaries and confirmations/contradictions).

[25] The distinctions made in Table 2.3 are only a compilation of existing descriptions and are certainly neither complete nor free of contradictions. The determination of text types must be regarded as an open process. However, the fact that the text types described here can actually be found is at least probable and has already been repeatedly demonstrated for some text types.

Table 2.3 Response text types, questions and expected effects on respondents. (A detailed explanation of the individual text types and question forms is given in Sect. 3.2)

Intended text type	Types of questions and speech prompts	Sample questions	Expected impact on respondents
Narration	Prompt to narrate: as a request or question (especially how- and what-questions about events and experiences)	Please tell us something about your school days! What was it like when you first got a job reference?	R. is motivated to narrate a story line or longer episodes
Description	Description prompt: as a request or question	Please describe your typical working day! What was your apartment like back then?	R. is motivated to describe experienced situations, places, processes, etc.
Justification/ explanation	1. Request for reasons (esp. why-questions) 2. Confrontation and polarisation: I. confronts R. with alternative option or presents alternative options	1. Why did you choose this course of study? 2. You have told that your studies were very strenuous. But otherwise your subject is considered rather easy. What do you think about that?	R. Is encouraged to explain or justify own actions and opinions
Reflection	Evaluative, normative and speculative ("hypothetical") questions	Should grades be used to measure performance?? How should studies be organised in order to be better motivated?	R. is encouraged to reflect and take a stand (e.g. on alternative ways of living and acting).
Information	(Follow-up) questions to ensure understanding (e.g. where, when and who questions)	When did you start your studies?	R. is prompted to clarify previous statements
Validation (and explication[a])	1. Reflection back: I. offers B. "spontaneous" interpretation or summary. 2. Conclusion: I. draws conclusions from R.'s answers and questions R. about them	1. First, let me get this straight: You didn't really want to study? 2. Were your difficulties in your studies a consequence of this unwillingness?	R. confirms the "offer" or explicates previous answers

(continued)

Table 2.3 (continued)

Intended text type	Types of questions and speech prompts	Sample questions	Expected impact on respondents
Explication	1. False summary: I. offers R. a "spontaneous" but deliberately incorrect summary or interpretation 2. Deliberate insinuation or suggestive question: I. asks a question which is recognisably suggestive for R.	(both forms cannot be presented here, as they can only be developed from the course of the interview)	R. is motivated to explicate previous answers

[a]Reflections and conclusions can lead to both validation and explication (see Witzel 1985, p. 246 ff.): If the interviewee refuses validation, this will usually lead him/her to (further) explications

For the discursive interview, it is decisive *which text types contain derivations* or which text types best enable their reconstruction. It is assumed here that this is the case above all with *justifications (explanations)* of one's own actions and judgements (or statements), because with these there is a particular need for plausibilisation and validation. This is based on a broad understanding of justification, which includes not only argumentations but also causal explanations (but not explanations in the sense of instructions). In contrast, vindications are not intended, although they are a probable response to justification requests.

Justifications (explanations) are therefore the central text type to be aimed for in the discursive interview. Therefore, speech prompts that are suitable for eliciting justifications are correspondingly important. In addition, other text types will also contain derivations (see Sect. 3.2). This can be expected in particular for reflections, validations and explications.

Data Collection with Discursive Interviews 3

The discursive interview is understood as an integrated research design in which all research phases (especially questioning, data collection and data evaluation) represent a conceptual unit. How this looks like for the Discursive Interview will be presented in this and the next chapter, first for the processes of data collection (Chap. 3) and then for the interpretative-reconstructive and further analyses (Chap. 4).

The order of presentation is oriented to the temporal (and to common forms of presentation). Nevertheless, it should not be forgotten that although data collection necessarily takes place before data interpretation, the procedure for data collection is conceptually derived from the requirements of data interpretation and analysis. For the Discursive Interview this means that the project of reconstructing collective representations from individual derivations within the framework of a contrasting strategy of interpretation requires certain forms of data collection: Both the sampling and the method of data collection must be oriented towards the goal of obtaining data material that is well suited for contrasting analyses.

Like many qualitative research designs, especially when they are designed to contrast, in the Discursive Interview the two main phases of data collection and data analysis take place as alternately as possible. The collection of initial data is followed by a phase of data analysis, on the basis of which a new decision is then made about further data collection. Even if the individual research phases are presented separately here, it should always be clear that the individual procedures of data collection and analysis are not self-contained and consecutive research steps, but should be carried out alternately and repeatedly.

Once the general methodological design (collective representations-analytical research question, contrastive analysis) have been defined, the search for data material necessary for answering the research question can begin. The totality of

C. G. Ullrich, *The Discursive Interview*,
https://doi.org/10.1007/978-3-658-38477-7_3

the procedures involved is usually referred to as data collection. Usually this consists of a two-stage process: first, the 'data subjects' must be identified and recruited (or made available) to participate in the research (sampling and recruitment). Subsequently, procedures must be applied to extract the necessary data from the data carriers (data collection in the narrower sense).

For the discursive interview search for data means: First of all, participants must be found and won over for taking part who can be assumed to be bearers or users of the collective representations that form the object of research (Sect. 3.1). These are then interviewed with discursive interviews in such a way that they reveal as many of their derivations as possible (Sects. 3.2 and 3.3).

3.1 Sampling and Recruitment

Most research designs require a selection of research participants (or units). Even an analysis of collective representations based on discursive interviews cannot include all potential bearers of collective representations. Therefore, it is necessary to make a selection of (types of) participants from the totality of all eligible participants (in the case of the Discursive Interview, these are all persons supposed to have research-relevant derivations), who are then interviewed.

The selection of respondents again takes place in a two-stage process: First, the criteria for participation must be defined. Once this has been done, people who meet the sampling criteria must be recruited to participate in the research.[1] Closely related to this is the question of how many research units (participants) are necessary.

While the first step, sampling, is a methodological question which is determined mainly by the general research interest, the so-called recruitment of participants who fulfil the sampling criteria is a practical research task, which can often enough come into conflict with the methodologically or theoretically derived sampling requirements – namely whenever it is not possible to persuade a sufficient number of persons who fulfil the sampling criteria to participate.

[1] Often enough, these two steps or aspects are not sufficiently distinguished and forms of recruitment are described as sampling strategies (cf. among others Merkens 1997; Przyborski and Wohlrab-Sahr 2014, p. 181 ff.; Reinders 2005, p. 120). More rarely, criteria for the selection of cases for (individual) case analyses are also counted among sampling strategies (Patton 1990).

3.1.1 Sampling

For the discursive interview, the core task or minimum requirement of sampling is to achieve that (as far as possible) only persons are interviewed who are relevant in the sense of the research design, i.e. for whom it can be assumed with a high probability that they will reveal individual derivations in the interview.

The question of determining the right units of study also includes the reverse question of how to prevent people or units that are not relevant to the planned research from nevertheless being sampled. In case of doubt, i.e. if there is uncertainty as to whether, for example, a type of person meets the sampling criteria, it is better to include it: The damage caused by "unsuitable" participants is certainly less than that caused by missing or omitting suitable ones.[2]

In contrast to the logic of quantitative sampling procedures, which is entirely oriented towards ensuring statistical representativeness, the goal of qualitative sampling is primarily to ensure good preconditions for empirically grounded theory formation in the sense of Glaser and Strauss (1967) – or more generally: for empirical-inductive strategies. The general goal of all qualitative sampling strategies can therefore be considered to be to find units of inquiry that contain the data suitable for this purpose in the most comprehensive and appropriate form possible. Accordingly, the study samples in qualitative social research are also not obtained as random samples, but by "purposeful sampling" (Patton 1990, p. 169 ff.), i.e. on the basis of targeted criteria derived from the research objectives and intended methods. This can have very different consequences, depending on the research question and research design.[3]

For contrasting analyses, the sampling should also be aimed at the maximum *variation of the objects of study* (persons interviewed), as this best ensures good contrasting opportunities and thus increases the chances of theory generation. With the systematic variation of the objects of investigation, the additional aim can be to cover the investigated subject area as completely as possible. In this case, the sampling goal is a categorical or *(typo)logical representation of* the investigated field,

[2] In the case of interviews, one would usually find out during the interview that the interviewee was wrongly recruited. This is certainly annoying because one has done useless work, but can hardly be avoided. However, it is much more problematic for gaining knowledge in the research process if, conversely, types of persons relevant to the research are not interviewed.

[3] For different qualitative sampling strategies see, among others, Kelle and Kluge (1999, p. 38 ff.), Merkens (1997), Patton (1990, p. 169 ff.) and Reinders (2005, p. 135 ff.).

which is a central prerequisite for type formation (cf. Sect. 4.4). Sampling goes beyond a "mere" systematic variation of the objects of investigation and aims at the collection of as many different forms of potential "data bearers" (users of derivations) as possible.[4]

For the analysis of collective representations with discursive interviews, the sampling task must therefore be enhanced: It does not only consist of determining the "right" persons for capturing collective representations; rather, an attempt should be made to include as many categories as possible of the "data bearers" relevant to the research question in the study sample. Such a (typo)logical representation can only be an ideal target. For, of course, one cannot know in advance whether and when a field is typologically represented. This can only become apparent in the course of the research and is itself an important research result.[5] Frequently, however, one will want to find good reasons in advance for at least individual cornerstones and take these into account in the sampling. Likewise, in an alternating approach, new categories can emerge in the research process and be used as criteria for further sampling.

(Typo)logical Representation

(Typo)logical representations should not be confused with statistical representativeness. Typological representation does not mean quantitative or distributional equivalence of sample and population, but rather the mostly equal "representation" of all empirical variations of a field in the study sample. This representation is thus completely independent of the empirical frequency and distribution of categories. It either takes place category-egalitarian or is weighted according to theoretical-conceptual considerations.

(*continued*)

[4] For both the systematic variation and the (typo)logical representation of the investigated field, a distinction must be made between research interest and the sampling strategy: The research interest consists in the greatest possible variation of the derivations of a reference problem. Sampling, however, takes place at the level of the units of study (participants). The only possibility to generate a high or complete variance of the derivations of a reference problem therefore consists here in the systematic variation of the units of investigation.

[5] Therefore, this (typo)logical representation in sampling must also be clearly distinguished from type formation as an analytical step in the evaluation of the data material (see Sect. 4.4), for which it is, however, an important prerequisite.

(continued)

1. Example of typological representation:
   In a study on statutory health insurants, which dealt, among other things, with the perception of (one's own) health insurance fund (Ullrich 2000), insured persons of all health insurance fund types (local health insurance funds, substitute health insurance funds, etc.) were surveyed with equal frequency in order to be able to capture differences between insured persons of different health insurance fund types as well as possible. Differences with regard to size (number of members) of the different health insurance fund types, on the other hand, which were considerable, were not taken into account in the sampling.
2. Example of typological representation:
   Students from two study programmes will be asked about their motives for studying. Based on preliminary assumptions, gender and the course of study are assumed to be important explanatory variables. The ratio of male to female students is 90:10 in course A, and 30:70 in course B. In the year under investigation, there are 800 students in course A, but only 150 students in course B. In a typological-egalitarian representation, these differences in frequency and distribution are ignored in each case. The same number of students would be surveyed in both study programmes, and the same number of female and male students in each case.

For achieving the sampling goals, qualitative social research can resort to two basic strategies that are opposed in their sampling logic: "theoretical sampling" and "selective sampling". Both can also be usefully employed in the reconstruction of collective representations on the basis of discursive interviews.

"*Theoretical sampling*" was introduced by Glaser and Strauss (1967, p. 45 ff.) within the framework of grounded theory and is probably (at least nominally) the most common sampling procedure in qualitative social research. Here, the selection of research participants is essentially based on initial findings from the current research process (e.g. selection of further interviewees only after and on the basis of the analysis of the first interviews). Theoretical sampling then leads to "alternating data acquisition": since the selection of new research units only takes place after an initial data collection phase and at least a review of the already gained data material, these three processes proceed partly in series, but partly also in parallel. However, there is no rule for the selection of the initial case.

In "*selective sampling*"[6] (Schatzman and Strauss 1973), on the other hand, the study units (participants) are selected on the basis of predefined criteria. These can be defined in different ways. If contrastive analysis is intended, these criteria will be based on characteristics that make a representation of the field – according to existing prior knowledge and always only provisionally – more likely. The aim here is to include all "theoretically relevant combinations of characteristics" (Kelle and Kluge 1999, p. 38).

"Theoretical sampling" and "selective sampling" follow opposite logics: Whereas in theoretical sampling the relevant selection criteria must first be obtained inductively, as it were, from the initial interpretations, selective sampling presupposes sufficient knowledge about the subject matter. Despite their opposites, both sampling strategies must be used in contrasting procedures. In this context, "theoretical sampling" functions as the basic form – on the one hand, because, as Glaser and Strauss (1967) explain in detail, this is the ideal procedure for empirically grounded theory generating, and on the other hand, because only in this way is it possible to openly (inductively) explore a field. Pre-determined selection criteria limit the horizon of knowledge, the possibility of gaining new insights, and this should not be done without careful consideration. However, this is necessary for the initial case selection which cannot be determined by "theoretical sampling"; here, content-related (theoretical) considerations are required.[7] Furthermore, it seems justifiable to use theory-guided aspects in a restrained form also in further sampling, especially if no or only a few selection criteria should result from the initial data interpretations.

3.1.2 Recruitment

In qualitative social research, recruitment refers to the concrete, practical acquisition of research participants (in the discursive interview: interviewees) on the basis of the applied sampling procedure. Thus, almost every research faces the problem of actually gaining the desired types of participants, especially those determined by "theoretical sampling". Provided that there are enough potential and equivalent participants who are also available for a study, the concrete decision as to which

[6] Selective sampling is often described in other terms (cf. among others Kelle and Kluge 1999; Patton 1990, p. 169 ff.; Reinders 2005, p. 136 ff.). For a detailed description of different forms of selective sampling, see especially Kelle and Kluge (1999, p. 46 ff.).

[7] Glaser and Strauss were already aware that pure "theoretical sampling" is not possible. They describe "theoretical sampling" as a procedure that begins after the "initial collection of data" (1967, p. 47).

persons (etc.) are actually included should be left to a random selection. Otherwise, there is a risk that "non-sampling" factors will have an undesirable and uncontrolled influence on the sampling.[8]

Often enough, however, it proves difficult to recruit the desired types of participants at all or in the intended number. There may be many reasons for this: the desired participants may simply be very rare (in the case of specific characteristics or combinations of characteristics) or hardly recognisable to the researchers, especially if they have only limited knowledge of the field under investigation or no access to it. Frequently, however, low accessibility of certain types of participants also proves to be an obstacle. This can be due to material/technical reasons (e.g. homeless people), but also to legal or institutional restrictions (e.g. inmates of closed institutions). A low willingness to participate on the part of specific participant types – in general (classic: self-employed) or with regard to the research topic (for example, in the case of "intimate" topics) – can also make recruitment considerably more difficult.

The problems that can arise during recruitment should not be underestimated. They can derail research at an early stage or make it appear impossible. Far more often, however, recruitment problems will researchers force to modify their research topic because it is not possible to sample in the intended way.

Given the frequent problems encountered in recruitment, it is not surprising that several techniques are widely used in qualitative social research to overcome potential barriers to recruiting participants.

Perhaps the most important technique is mediation by *gatekeepers*. Gatekeepers are persons who, due to their proximity to the field of interest (or a special field access), are able to establish contacts with potential participants. Very often these are members of certain professional groups (e.g. teachers, social workers, physician) and social groupings (e.g. of ethnic groups or subcultures) or employees with managerial functions.

[8] Such "random recruitment" is not to be confused with the random sampling of standardized research and also only takes place within or below "theoretical" or "selective sampling". Its aim is not representativeness, but the avoidance of arbitrariness. The most important "non-sampling" factors include the spatial proximity to the researchers (short journey = higher probability to be recruited), the situational accessibility of potential participants (e.g. whoever is contacted first by telephone participates), personality characteristics of the potential participants and their esteem by the researchers (e.g. those who are sympathetic to the researcher are more likely to be asked to participate) or just sequencing effects (e.g. alphabetical order of potential participants).

The *self-selection of* participants is another widespread procedure. Various forms are used (classic: newspaper advertisements, postings in central places, calls in the local press, etc.; today often also via internet media). The participants have to enlist themselves. In order to overcome this possible obstacle to participation, expense allowances are usually offered.

A third possibility is the *snowball method* (sometimes also chain method). This consists of asking the participants who have already taken part to recommend other participants. This can be successful because the research participants often have better knowledge of the field of interest and often have contacts with people who also meet the sampling criteria.

Due to the difficulties that can arise in obtaining participants, one will often enough have to resort to such strategies of recruitment in research practice. However, this should be limited to well-justified exceptions. This is because all three recruitment channels imply that researchers loose a (possibly even considerable) part of their control over the sampling process. This is problematic because these ways of recruitment can lead to biases of various kinds.

Problems of Different Recruitment Channels

Gatekeepers often have their own interests or misunderstand the sampling criteria given to them; and of course gatekeepers can only know "their" people. It is all too easy to be mistaken about whether and to what extent gatekeepers actually have special field access (both the researchers and the gatekeepers themselves). Moreover, the choice of gatekeepers is itself a selection process, so bias can occur at two stages here: In the decision about the gatekeepers and in the mediation of contacts by the gatekeepers.

Self-selection often does not lead to good recruitment because the incentive to enlist oneself is too low. If this is increased by financial compensation, this can in turn lead to false incentives and a corresponding distortion, e.g. if readers of an advertisement "promotes" this information (financial compensation) intensively in their own social context. In addition, it is usually not possible to ensure that all eligible participants are reached with self-advertisement. Therefore, this procedure also depends on coincidences of awareness (e.g. whether people read a newspaper advertisement) or favours certain types of recipients.

Similar problems arise with the *snowball procedure.* Here, the participants already studied have a similar effect to the gatekeepers; above all, however, the participants can usually only recommend or request people from social context. Similar to self-selection, there is therefore a risk of bias that cannot be identified and controlled.

3.1.3 Sample Size

In addition to the rules for sampling and for the recruitment of participants, the size of the sample has to be defined. At the outset of the research, there cannot be fixed rules and a strict determination. This would contradict the logic of qualitative research designs too much. The handling of the sample size is largely determined by the research design (from case studies to extensive comparisons) and the sampling strategy derived from it.

In "theoretical sampling", the sample increases incrementally with the decisions made in the research process about new criteria for the selection of cases. Without a stopping rule, the process of theoretical sampling could in principle be continued indefinitely – at least as long as reasons for new selection criteria are found. Precisely this, however, already conceals the usual rule for bringing the sampling process to an end.

Here, Glaser and Strauss (1967) introduce the criterion of "*theoretical saturation*": research units are as long selected as this is considered useful for the research. If, in contrast, a "saturation effect" occurs in the process of data interpretation, i.e. the feeling that no significant new insights can be gained through further data (ideally, because the empirically grounded theory formation is already far advanced), the sampling can be terminated immediately.[9]

The fact that it is not yet possible to provide precise information on the number and type of research units at the planning stage often creates uncertainty (especially when research experience is still limited). However, this should be tolerated with regard to the research objectives and an optimised approach.

In any case, theoretical sampling in its pure form is often difficult to implement in research practice, especially because the funding of research projects depends to a considerable extent on the sample size. Here, but also, for example, in the case of student theses, more precise information is often requested in the planning stage. It is therefore important and generally possible to develop realistic scenarios about the type and size of the sample. Anyhow, in case of "selective sampling" it must be decided at the outset which and how many types of research units are necessary or desirable for the research. How many units are needed for each type (e.g. how many respondents with similar characteristics) cannot, however, be precisely determined in advance.

[9] Lincoln and Guba (1985) speak of (informational) redundancy in this context: "If the purpose is to maximize information, the sampling is terminated when no new information is forthcoming from new sampled units; so redundancy is the primary criterion" (Lincoln and Guba 1985, p. 202).

Finally, the size of the sample (or the estimate of how many cases are required for "theoretical saturation") is also determined by some practical research factors, such as the methods used and their workload, as well as, of course, the resources available (time, number of researchers). Like other contrasting methods, an analysis of collective representations with discursive interviews requires a relatively large sample, because this is the best way to ensure a sufficient amount of contrasting options. However, a further delimiting of the sample size is neither possible nor necessary. It will become apparent in the process of "theoretical sampling" and saturation.

3.2 The Discursive Interview as a Guided Interview

As was made clear in the second chapter, the discursive interview aims to obtain text types that contain individual derivations. The discursive interview makes use of the fact that collective representations have to be communicated and interactively validated on the one hand and that they are necessary on the other hand for explaining and justifying decisions and opinions to others. By means of various questioning techniques, interviewees should be encouraged to give respective answers (especially explanations of their decisions and opinions).

In order to ensure this well, discursive interviews are conducted dialogically and guided. What guided interviews are, why the discursive interview is designed as a guided interview and what types of questions are included will be explained in this section. The following Sect. 3.3 deals with particular aspects of discursive interviewing.

3.2.1 Guided Interviews

Guided interviews are probably the most frequently used interview technique in qualitative research. However, it is not entirely clear what the term guided interview actually means. On the one hand, guided interviews can be regarded as an independent type of interview and distinguished from other interview forms (e.g. Misoch 2015; Przyborski and Wohlrab-Sahr 2014). On the other hand, one can also speak of guided interviews in the sense of a generic term and subsume under this various interview forms in which guides are used (e.g. Flick 1995; Hopf 1988).[10]

[10] It is also striking how often the term "guideline interview" is not defined in more detail. This is particularly irritating where the practical use of guidelines in research is a central topic (e.g. Helfferich 2009; Reinders 2005).

In both variants, the classifying use of the term guided interview is unfortunate, because it makes a technical feature to a fundamental distinguishing criterion. It is not very useful for denoting as an interview type, because this interview type would not be internally consistent and cannot be meaningfully delimited externally; and as a generic term, "guided interview" designates an extremely heterogeneous group of interview forms that converge only in this one, subordinate aspect.

Nevertheless, the term guided interview can be used to characterise interview forms – as one among several possibilities (such as "expert interview" or "online interview"). However, the use of a guideline alone is not sufficient for this, as this is common in almost all qualitative interview forms. The decisive difference is rather the function of a guideline in the interview: Only if the guide is used to structure the course of the conversation should this be called a guided interview, but not if guidelines are only supplementary or subsidiary to other forms of communication.

Basically, the use of guidelines can be seen as the consequence and attempted solution of the tension between the need to structure the interviews on the one hand and the interest in as much openness and spontaneity as possible on the other. Whether and how guides are used must and can in fact only result from methodological considerations and from the research interest.

The reasons for structuring interviews with a guideline can vary in detail. In essence, it is usually a matter of ensuring that all topics considered relevant to the research question are actually the subject of as many interviews as possible (or that at least in all interviews the opportunity for this is given), even if their thematisation should not "automatically" emerge from the interview interaction.

This interest can result from different factors and will often enough derive from theoretical considerations or "deductive" needs for the thematisation of certain contents (so e.g. Reinders 2005, p. 135 f.). This also includes more intuitive, common sense theories, especially about the improbability of certain topics becoming the subject of an interview (Helfferich 2009, p. 178 f.). However, such an approach would not be compatible with the reconstructive design of collective representations analysis approaches.

However, the use of guided interviews is often justified by the interest in a thematic comparability of the interviews. Thus, a minimum of consistency of central topics is a necessary prerequisite for contrasting analyses. Finally, the interest in certain types of questions (and prompts) and answers related to them or the use of technical aids (e.g. photos, vignettes) can, if not require, at least suggest a guideline structure.

For the Discursive Interview, two aspects are crucial for a guided interview:

1. On the one hand, there is an interest in a more dialogical form of conversation with a clear distribution of roles (interviewer asks, respondent answers). In principle, this would also be possible without the use of guidelines. However, since in the discursive interview a number of specific (and in many cases only optional) *forms of questions and interview techniques* are used, it is advisable to use an elaborate guide.
2. However, guideline support in the Discursive Interview is finally necessary for ensuring the *comparability of* the interviews for the *contrasting analyses.* For, according to a central assumption of the Discursive Interview (see Chap. 2), the reconstruction of *collective* representations is only possible through systematic comparisons of the individual derivations revealed in the interviews. In order to reach this comparability, the interviewers must be able to direct the conversation towards the reference problems relevant to the research, should this prove necessary.

Conversely, a guided interview means that the further interaction context of utterances is of rather little importance for the reconstruction of collective representations. Above all, however, the preservation or elaboration of a "case structure" (in the sense of an interview transcript or even the interviewee) is not necessary. The interviewees are primarily of interest as users (reproducers as well as shapers) of collective representations.

3.2.2 General Criteria for Constructing an Interview Guide

Overall, there is a large discrepancy between the frequent use of guided interviews and a lack of methodological reflection. Even more extensive textbooks (Helfferich 2009; Misoch 2015; Reinders 2005) are almost exclusively limited to practical research questions, which in turn are dominated by or even exclusively deal with the practical problems of interviewing.

The insufficient methodological basis becomes most obvious in the planning and construction of the interview. For many decisions– whether a guide should be used at all, what form a guide, how it should be used in the interview, and how to construct a guide – there are often no suggestions or instructions at all, but if there are, they are based on common sense views and the authors' research experiences.

This lack of reflection is certainly also due to the fact that guided interviews are not a homogeneous group of methods, and that the question of which form a guide should take cannot be judged in isolation from the respective methodological ideas and research interest. Nevertheless, some basic rules can be formulated for the construction of guides, albeit only in general terms and based on some general considerations. These basic rules for construction interview guides will be briefly outlined below, before the more specific techniques of the Discursive Interview are explained in the next sections.

The construction of a guideline, at least a relatively structured one, is much more complex than usually expected at the beginning of the research process. When constructing a guideline, a number of aspects have to be taken into account, which can lead to various trade-offs. Above all, however, there are four basic aspects or goal functions to consider when developing an interview guide:

1. Why is the question asked (or the question-equivalent stimulus given)[11]? Actually, this should be clear in any case. After all, apart from functional questions (e.g. filter questions), which also occur in qualitative interviews, for all questions of an interview guide their importance for the general research interest is of course decisive. Nevertheless, practical experience shows time and again that the seemingly simple point of why a guideline question is asked often cannot be explained, or only in a very vague way.

   An important reason for this often observed difficulty can already be assumed in the principle of openness or in the explorative character of qualitative research. After all, even with guided interviews, no simple "data" is sought to test hypotheses. Rather, an attempt is also made to gain new, not anticipated insights. This means that questions can often have a "speculative" value; their usefulness cannot be known for sure in advance. However, a plausibility check can be expected here, namely good reasons for the "speculation" that a question will lead to answers that *may be* relevant in terms of the research question.[12] In this sense, questions can also be compared and assessed in terms of their importance for the general research interest.

[11] When "questions" are referred to in the following, this is not meant in the literal sense, but as defined in Sect. 2.3, namely as information-generating question-shaped or -similar speech prompts. "Stimulus" is used here as an alternative to "prompt to speak".

[12] However, this is then also the minimum requirement for a guideline question. If it is not met, this leads to both practical and research ethics problems. While the former is merely annoying (a superfluous question is asked that eats up time for others), the latter is a serious problem: Questions that cannot be justified by research interest have no place in a guide. To ask them anyway would be pure voyeurism.

2. What is asked about?
   If the (presumed) theoretical relevance of a guideline question is not clear, this may also be due to the fact that the "target field" of possible answers is blurred. For closely connected with the theoretical relevance in the sense of the question is the *content dimension* of a question. For as long as it is not clarified how a question can be answered, which types and dimensions of answers are possible, it is also not certain whether or to what extent research-relevant text material can be expected in the answers. Thus, while the first aspect should ask "Why do I want to know this?" the question here is "Is what I am asking about what I want to know?" It is all too easy for the expected answer(s) to differ from the actual ones. The spectrum of possible answers should therefore be anticipated ("played through" in advance) as well as possible. In this way, it can be determined by means of thought experiments or a kind of "pretest" (mock interviews) whether a question actually aims at the intended content or how precisely it does so (see also Sect. 3.3.2).
3. Why is the question phrased this way (and not differently)?
   In addition to the determination of the content, its formulation is also fundamental for the quality of a question. Here, it must first be clarified which stimulus or question type (see below) is appropriate. In a second step, the appropriate formulation should be selected by comparing and trying out ("playing through") different formulation alternatives, whereby, in addition to the comprehensibility and unambiguity of the formulation, the expected "fruitfulness" (strength of the response stimulus) as well as the type of response intended by the question (text type) are the most important criteria.[13] (Further aspects of the formulation of questions and question-like prompts are discussed in detail in Sects. 3.2.3 and 3.2.4).
4. Why is the question or group of questions at this point in the guide?
   This is about the *coarse and fine structure of a guideline*, i.e. the order of questions or groups of connected questions. Ideally, it should be possible to clear for each question why it is in a particular place in the guide. Often, this is due to obvi-

[13] The formulation aspect is not invalidated by choosing a form of guideline that does not require the guideline questions to be completely formulated. The advantage of no or only vague formulations in the guide is usually seen in the fact that this allows the interviewer more spontaneity and flexibility. However, there is a very high risk that spontaneously asked questions miss the intended formulation and are therefore misleading. Thus, at least for discursive interviews, it is by no means irrelevant to ask a why-question instead of a how-question, because in this way fundamentally different types of answers are intended or generated. For this reason, even a partial renunciation of formulated questions must be accompanied by a correspondingly high level of awareness of the intended type of answer.

ous content-related reasons (e.g. the necessity of a certain order of topics) or technical reasons. The same applies to the distribution of question types in the guide. For example, it may be advisable to use certain types of question (see in particular Sect. 3.2.4) only at a relatively late stage in the interview, because respondents may feel under pressure to justify themselves and therefore reduce their willingness to talk. Furthermore, the *relationship between single questions* is closely related to the sequential structure of a guideline. Particular attention should be paid to the "hierarchy" of questions (e.g. obligatory main questions and optional follow-up questions that depend on the answer to the main question).

Ideally, solutions should be found for all issues arising from these four aspects in interview guide construction.

Aspects of Interview Organisation in the Guided Interview

Beyond obtaining meaningful answers, questions in the guide can have different tasks that can be of great importance for the handling of the guide or for a successful interview. With regard to the general conduct of the interview, several types and functions of questions can be distinguished:

1. **Information or filter questions**:
   Information or filter questions target singular pieces of information that are easy for respondents to provide (such as a person's age or occupation). Their purpose is usually to ensure that the correct parts of the guide are used. This is because often not all questions in the guide are appropriate for every respondent. If the relevant information is not already known before the interview, it may have to be obtained during the interview. In addition to this control function, information questions can also be important for later interpretative analysis.[14]

(*continued*)

[14] Such information is usually collected after the interview, e.g. using a short questionnaire. However, a need for information often arises during the interview. Information questions are then usually asked as spontaneous follow-up questions. Since this can very often become necessary, it should already be considered in advance how, when and to what extent information questions can be asked.

(continued)

2. **Main questions and dependent questions**:
   Many questions can (and usually should) be asked all respondents (main questions). In addition, there are often guideline questions that are dependent on the course of the interview or on the answer to a previous question, i.e. can only be asked if the answer to the main question suggests this. Accordingly, the guideline questions are often "logically hierarchised". This does not mean, however, that follow-up questions are also subordinate or optional in the sense of being an additional possibility. On the contrary, these follow-up questions can often be central to the investigation. Therefore, when developing questions, attention should be paid to what conditions must be met in order for a question to be meaningful. The complicating factor here is that several parallel "follow-up questions paths" may often be necessary.
3. **Conversation Framing:**
   This includes all questions or activities aimed at defining the situation "qualitative research interview" before and during the interview – and in a way that supports the research interest. In particular, these are measures to create a constructive atmosphere for the interview and to increase the interviewees' competence and willingness to talk (including informal small talk, active listening), informational talk (including information about the research interest, clarification of the role of the interviewee) as well as framing and facilitating the start of the interview (including warm-up questions) and the end of the interview (conclusion, résumé, etc.).
4. **"Revisits":**
   It is often useful to revisit a topic at a later point in the interview. Such revisits are appropriate, for example, when a topic is to be addressed in different contexts. Revisits can be planned only to a very limited extent. Frequently, however, thematic breaks, repetitions or cross-connections cannot be avoided during the construction of the interview guide. If this is recognised, it is certainly advisable to consider how a topic can be taken up again appropriately.

3.2.3 Question and Stimulus Types

In terms of form and intended effect, a variety of different questions and question-like prompts can be distinguished. It is therefore of fundamental importance for the construction of a guideline to be clear about how the formal structure of questions *can* affect the response behaviour of the interviewees, i.e. what type of response is *suggested* by a question wording.[15] In the following, the most important question types are distinguished on the basis of the text types they intend (cf. Sect. 2.3.2; Table 2.3) . Finally, in a further Sect. 3.2.4 specific stimuli are presented which are of particular importance for the discursive interview.

With regard to the intended text types, five basic, but by no means mutually exclusive, question types can be distinguished:

1. **Comprehension and knowledge questions:**
   Not only are comprehension and knowledge questions often indispensable for handling the interview guide; there is often likely to be a substantive interest in respondents' knowledge as well. On the one hand, it is easy to ask about the knowledge or expertise of interviewees. However, knowledge questions can have a detrimental effect on the interview situation, for example if interviewees feel they are being interrogated, if the interview situation feels like an oral examination or if interviewees are given the feeling that they are incompetent. In principle, it must be assumed that knowledge deficits of the interviewee that become clearly apparent in the interview can lead to a deterioration of the interview atmosphere and to a reduction in the willingness to talk. Therefore, if there is a strong interest in the knowledge of the interviewee, it is probably more advisable to identify knowledge stocks as indirectly as possible. Another means of avoiding negative effects is subjectivising "assessment questions", which are formulated in such a way that the possibility of a "wrong" answer basically does not exist (e.g.: "What do you understand by [...]?"). As a matter of principle, the impression should never be given that the respondents' statements are false. They should therefore only be corrected if this is indispensable for the progress of the interview. In the context of discursive interviews, comprehension and knowledge questions are of little importance and are only used to a limited extent.

[15] There is, of course, no deterministic relationship between questions and answers. No question "forces" a certain form of answer. If, for example, an explanation (cause-effect relationship) is asked, it is absolutely possible to answer with a narrative or a justification.

2. **Narrative and descriptive prompts:**
   Narrative prompts are all stimuli that aim at a narration of past events, experiences and actions. They form the core of most qualitative interview procedures. Narratives generated in interviews are also quite important for the Discursive Interview, because narratives also draw on collective representations.

   Nevertheless, the Discursive Interview aims primarily at justifications and explanations as a text type (TT) or justification texts as a text type element (TTE), because these form the best source material for the reconstruction of collective representations. The immediate importance of narrative texts for the reconstruction of collective representations is thus limited, although it should be noted that narratives (TT) usually also contain justification texts (TTE). The importance of narrative prompts in the Discursive Interview therefore lies less in the fact that collective representations are already evident in narrations, but rather in the fact that narrations are an important prerequisite for questions directly aimed at eliciting derivations: Only when sufficient "narrative material" is available can stimulus forms aimed at justification (or explanation) be used.

   If the direct significance of narrations for the recording of individual derivations is therefore secondary to that of other forms of questioning (because narrations per se contain rather few derivations), their indirect significance in the Discursive Interview is all the greater. This is because both requests for justification and many explication and validation questions presuppose a link to narrations.

   Similar to narrations, descriptions (especially of situations) can also contain individual derivations (e.g. stereotypes). Therefore, they can also be used for an analysis of collective representations. Descriptions, however, differ from narratives in that they usually do not allow the connection of requests for justification or explanation. Therefore, descriptions are of little importance for the discursive interview in comparison to narrations; accordingly, requests for descriptions should be rare there.
3. **Requests for justification:**
   Requests for justification (or explanation) can refer equally to narrations of past actions and situation definitions, to current situation definitions, assessments and action orientations, and to intentions for action expressed in the interview. They always presuppose a corresponding reference (something that can be justified or explained) and must therefore (usually) arise from the interview interaction. Their predictability is therefore limited, but can be increased by specific forms of questioning (see Sect. 3.3.4).

In most qualitative interview methods, requests for reasons are excluded, not explicitly provided for, or play a subordinate role.[16] For the discursive interview, on the other hand, requests for justification and explanation are of decisive importance, since it is assumed here that collective representations resp. derivations are expressed most clearly in justifications. Therefore, interviewers should to be able to generate justifications directly in case they do not yield from narrations and their follow-up questions.

The typical form of a request for reasons is the open-ended why-question ("Why do you think that [...]", "Why did you [...] at that time?" etc.). In contrast, offering possible reasons is problematic even if respondents do not initially "think of" any reasons. This is because the appeal of external offers of interpretation for one's own actions or their representation is likely to be very strong in many interview situations and can lead to a distorting tendency to accept such proposals for reasons.

Requests for justification (or explanation) can also have a detrimental effect on the interview situation. Particularly if they are supported by confrontation and polarisation (see Sect. 3.2.3), the interviewees could get the impression that they have to justify themselves to the interviewers. It should be clear that this would have an effect on the atmosphere of the interview and in the worst case could even lead to a break-off of the interview. But even if this does not happen, such an interview structure hinders the most open and broadest possible recording of individual derivations.

Though "social desirability" is not an immediate problem for discursive interviews, which are not interested in the "true" motives for action of the interviewees and the derivation actually effective (cf. Chap. 2), but rather want to capture and reconstruct communicable – and therefore social and at least apparently and/or in certain social areas desirable – collective representations. But, there is a danger here that only certain collective representations will become clear, namely precisely those that respondents assume the interviewer shares or are known or understandable to him/her. In other words, what is considered communicable by the interviewees is also determined by the respective interview situation and thus to a large extent by the person primarily addressed (usually the interviewer). What is considered communicable can change rapidly with the interview situation and the interview participants. The possible effects

[16] Exceptions here are the problem-centered interview according to Witzel (1985) and the semi-structured interview according to Scheele and Groeben (1988), which explicitly provide for requests for reasons.

of the interview situation must therefore also be taken into account when analysing the data material (cf. Chap. 4).[17]

These problems should be prevented by conducting the interview appropriately. Therefore, requests for reasons should only be made after a constructive and trusting interview atmosphere has been created. In addition, requests for reasons should be used sparingly so as not to get into a continuous discussion with the interviewees and thereby cause a preponderance of abstract argumentation. Requests for justification should therefore, if possible, only be made after longer narrative passages and refer to these. Such restraint or contextualisation (linking the request for justification to the interviewees' narratives) makes it possible to largely dispense with context-free requests for justification (explanation) and is less stressful for the interview situation.

4. **Requests for explication and validation:**
   In the discursive interview, questions and stimuli aimed at validation and explication can play a particular role. In the case of explications, interviewees are asked for additional explanations. Typical for this are follow-up questions of various kinds. These also include clarifications of actual or apparent contradictions. Questions aiming at explications are thus part of the basic equipment of all qualitative interview forms; in the discursive interview they are also to be used specifically for the elicitation of individual derivations.

   Validations go one step further than explications: Here, respondents are offered summaries and interpretations of what they have said, on which they can then comment again (for suitable interview and questioning techniques, see Sect. 3.2.4). Both forms of questioning necessarily follow on from justifications (more rarely also from narrations). They deepen a previously discussed topic. For the discursive interview, both forms of questioning offer a suitable way of prompting respondents to provide additional and more detailed explanations and reasons (TTE) and thus increase the chance of capturing derivations.
5. **Requests for evaluation and normative questions:**
   Finally, for discursive interviews, questions are of importance that directly encourage the respondents to take a position on certain facts, because here the probability of recourse to collective representations is very high. Such requests for statements explicitly aim at evaluations of situations, reference problems and actions. They can either follow on from the respondent's comments or be independent of them. With evaluative questions, it should be noted that they can

[17] The general importance of the interview situation and, above all, the effect of the interview interaction on data collection has already been pointed out in Sect. 2.2.

be generated in different ways and that this can have a significant impact on responses. Thus, evaluative and normative questions can be formulated with different degrees of openness, be very direct or rather indirect, and refer to general phenomena or to concrete life events of the respondents.[18]

Discursive Interview and "Natural Conversation"

What is often perceived as irritating about the discursive interview is that the interviews are relatively strongly structured and questions are asked rather "bluntly". This contradicts a widespread and powerful "naturalistic" philosophy of qualitative methods, according to which interviews should be conducted in a "natural", "egalitarian", "context-rich" etc. way. This "naturalism" in turn results from the idealisation of interaction models in the tradition of Goffman and Mead, which defines non-institutionalized face-to-face interaction as generic and prototypical, whereby other forms of communication such as telephone calls and correspondence (no face-to-face communication) or even job interviews and interrogations (institutionally regulated) almost automatically appear deficient.

The discursive interview, on the other hand, is based on an understanding of interaction that does not recognise forms of interaction that are not shaped by social norms and rules. Like all forms of interaction, discursive interviews are therefore institutionally shaped and just as little "natural" as all other forms of qualitative interviews. They are a specific form of communication that should first of all be unproblematic for all those voluntarily participating in it. Nevertheless, the Discursive Interview certainly differs from many other interviews methods in that it is conducted intensively, with the aim of motivating the interviewees to give reasons. However, such "discursive" forms of conversation can also be found in supposedly "natural" everyday communication. In this respect, the discursive interview is perhaps even more "natural" than many other types qualitative interviewing.

[18] Often, the interviewees' judgements will be based on evaluation criteria that were actually or supposedly specified by the interviewers. When asking for comments, an attempt should therefore be made to refrain from specifying evaluation criteria if these cannot be derived directly from previous statements made by the interviewees. If this is not possible, general terms should be used wherever possible (e.g. "good/bad" instead of "fair/unfair"). In addition, an attempt should be made to have the interviewees explain their understanding of the evaluation standards (e.g.: "What do you mean when you describe [...] as 'just'?").

3.2.4 Special Questions and Stimuli to Elicit Individual Derivations

Due to the high importance of justifications and other text types for the reconstruction and analysis of collective representations, an interview procedure used for eliciting individual derivations must be able to encourage answers that contain individual derivations to a higher degree. In the discursive interview, specific types of questions and stimuli are provided for this purpose, which go beyond simple requests for justifications, validations and explications and can be used to elicit specific types of answers:

1. **Confrontations and polarizations**:
   Confrontation questions are a relatively strong intervention in the conversation situation. Two basic forms can be distinguished. In the case of *"internal" confrontations,* interviewees are made aware of inconsistencies and contradictions in their statements and asked for clarification or for further or renewed explanations (TT). Internal confrontations are therefore always based on previous statements made by the interviewee.
   In *"external" confrontation*, on the other hand, the interviewees are confronted with alternative points of view, i.e. with attitudes, behaviours and their justifications that more or less clearly contradict the derivations already shown in the interview. External confrontations enable the interviewer to purposefully and discursively introduce theoretically relevant aspects that would threaten to remain untouched in the course of the interview without such intervention. External confrontations thus lead to an expansion of the argumentative spectrum. They are therefore particularly suitable for checking the stability of judgments and their justifications.
   Like external confrontations, *polarizations* also aim to expand the discursive potential in the interview situation. The difference to external confrontations is that here this is no longer done directly from the interview interaction. Rather, the interviewees are confronted with two or more contrary views of an issue on which they have not previously expressed an opinion, and are asked to make an statement. The most fruitful approach is certainly a smooth transition from external confrontations to polarizations, so that the respondents are first confronted with an alternative interpretation or evaluation, which is then successively supplemented by others. A completely unmediated offer of polarizations should be the exception and be limited to situations in which the interviewees cannot otherwise be motivated to evaluations .

In the case of external confrontations and polarisation, there is a danger that the alternatives offered will appear to be the opinion of the interviewer or to have some other special dignity (in particular as "prevailing opinion" or as "scientifically founded"). This danger can at least be reduced by a "distanced reference" of the contrary collective representations (e.g.: "But there is also the view that [...]. What do you think of this?/What would you object to this view?"). At the very least, "external" views should be avoided appearing as the interviewer's opinion. Rather, they should be formulated and offered as familiar and common views, not as dominant or even correct ones.[19] Polarizations should also be thoroughly examined and pre-formulated as much as possible. It can also be useful to "borrow" (cite) opinions and formulations from research already underway (interviews already conducted).

2. **Summaries and conclusions:**
   Another means of eliciting derivations are summaries and conclusions, both of which can be used primarily for validations and explications. Both summaries and conclusions are based on the already obtained narratives and justifications of the interviewees and presuppose situational "spontaneous" interpretations by the interviewers. They can be formulated in a purely "factual" manner or contain additionally evaluative elements.

   Summaries (or offered recapitulations) are more or less strongly pointed paraphrases of interviewees' narrations and justifications (e.g.: "So, if I have understood this correctly now, you have [...] because/about [...]"). *Conclusions*, on the other hand, are conclusions and exaggerations based on summaries (e.g.: "You told me that [...]. Could one say that you [...]?"). Conclusions thus consist of offers of interpretation on which respondents can then comment.

   For both summaries and conclusions, a distinction can be made between "correct" and "incorrect" ones. "Correct" summaries and conclusions are primarily aimed at validation and are thus "honest" offers of interpretation. This can lead to them being readily and possibly hastily accepted by respondents. This would be unproductive for the elicitation of derivations and can only be broken up by subsequent additions and differentiations by the interviewers.

   Derivations are easier gained with "false" or exaggerated conclusions and summaries. Similar to leading questions (see below), they encourage the respondents to modify their previous explications and thus to provide further ex-

[19] There are various possible formulations for this, e.g.: "Frequently/ in the media/[...] it is also claimed [...]", "But could one not also say/hold the opinion that [...]" or also "Other participants in the study have assessed [...] quite differently, namely: [...]".

planations and reasons. However, they must not be clearly "wrong"; in this case, they would only give the impression that the interviewer did not listen properly. Rather, they are over-pointed or highly simplified summaries and conclusions that do not completely miss the point of what was said, and thus can encourage respondents to provide additional explanations and to reveal more derivations.

3. **Deliberately leading questions and premises:**
   Even false summaries and conclusions have a "suggestive" character (even if easily recognizable). Moreover, leading questions and premises can also be used productively to elicit derivations.[20] Provided that this is done consciously, informed and purposefully, they can stimulate the respondents to further clarifications, explications and justifications.

   On the other hand, it can have a very detrimental effect on the interview situation and the quality of the answers if leading questions and premises "just happen" without the interviewer intending this or even noticing it. This can give rise to the suspicion among the interviewees that they are being pressured into certain answers (i.e. they are being "suggestively manipulated"). If a purposeful use of leading questions can also be assumed to be more sensitive to the dangers involved, the same problems can arise as with non-intentional leading questions and premises. Therefore, it should be noted above all that the willingness of respondents to be explicative is limited and that their ability to deal with leading questions can vary greatly from one individual to another.
4. **Hypothetical situations:**
   A further means of eliciting individual derivations are hypothetical situations, which are primarily aimed at the text type reflection. Here, the respondents are asked to put themselves in other situations and to consider what decisions, judgements, etc. they would come to in such a case ("If you were […], what would you do then?", "Imagine […]. How would you judge your situation then?" etc.). Through hypothetical situations, it is possible for interviewees to position themselves even on those facts of which they are not directly affected.

[20] As mentioned in Sect. 2.3, Richardson and his colleagues (Richardson et al. 1965, 1979) studied the effect of leading questions and found no problematic effects, provided that they were "informed". Despite the already classic status of Richardson et al.'s (1979) analyses, they have been incompletely recognized in qualitative research. Reinders (2005, p. 141) and Kruse (2014, p. 218), for example, adhere unwaveringly to the classic view of survey research and warn of the danger of leading questions. In contrast, Gläser and Laudel (2009, p. 132 ff.) recommend leading questions and premises as (very) useful, also invoking (in this simplification: wrongly) Richardson et al. (1979). Others, such as Hopf (1978) and Helfferich (2009, p. 106), refer to Richardson et al. (1979), but without adopting their justifications and differentiations.

In addition, individual parameters of an issue discussed in the interview can be varied. In this sense, hypothetical situations are usually linked to concrete, already discussed issues, which are then varied in thought experiments.

Hypothetical situations are also possible without a direct reference to interaction. For example, respondents can be asked to put themselves in the unreal situation of a person in positions of responsibility (e.g. a political representative or employer) ("If you had to make a decision as a [...]: What would you change regarding [...]?" etc.). Such rather unmediated hypothetical situations, which can be well supported with prepared material such as photos or vignettes, also provide information about basic derivations of the interviewees. The associated change of perspective, however, allows contradictions and inconsistencies in the explanations to emerge more clearly, which in turn can be used as internal confrontations (see above) to clarify and elicit further answers containing derivations.

Leading Questions and Premises

Overall, the idea that leading questions are to be avoided and represent an interview mistake also dominates in qualitative interview research. This is based on the belief that leading questions and premise either unintentionally distort the answers of interviewees or even deliberately manipulate them. Accordingly, both terms have quite negative connotations. Both forms narrow the possible range of answers of the interviewee in an inadmissible way through specific framing. Leading questions do this in an active form, in particular through a specific ("suggestive") form of wording. Premises, on the other hand, have an indirect effect through "omission", especially through non-explicit background assumptions that are treated as shared or unquestionable.[21]

However, the distinction between form and effect also applies to leading questions and premises(see Sect. 2.2): they can be "suggestive" or "insinuating" in purely formal terms or in their intended or actual effect. For example,

(continued)

[21] The following question can be taken as an example of both forms: "Unemployment is often associated with considerable financial and social problems for those affected. What could or should be done about it?". The – here open and clearly visible – framing "suggests" a problem that the interviewee (otherwise) may not feel at all. At the same time, the question already "implies" that there is not only a need for action (it is not asked "if"), but also possibilities for action.

(continued)

everyday communication has a high proportion of formally suggestive and insinuating formulations, without corresponding effects being observed or this being problematic for the addressees. Rather, it seems that (adult) persons can usually distinguish relatively easily between purely formal leading questions and premises and manipulative ones. Moreover, ("correct") insinuations are often unavoidable.[22]

The discursive interview takes up this everyday communicative use of formal leading questions. The guiding idea here is that formal leading questions can be used primarily to obtain explanations (but also justifications). However, this is only possible if leading questions are, on the one hand, consciously (purposefully) and informed used – i.e. mostly: if they are developed from the context of the interview – and if, on the other hand, the interviewees can therefore easily recognize the "suggestive" character of the question and react accordingly. Or in short: if questions are formally suggestive, but do not have a corresponding effect.[23]

With these general questioning techniques described here, which could be both supplemented and further differentiated, the discursive interview has a wide range of instruments at its disposal for generating justifications, explanations, validations, explications and reflections, and thus of the text types which can be assumed to contain derivations to a greater extent. Different text type elements can be generated (see Table 3.1). On both levels (text type and text type element), however, justifications and explanations are the forms primarily aimed at.

[22] Even such a simple question as "What is your name?" is based on several ideas and norms that remain implicit but are nevertheless assumed to be shared (among other things, that the person addressed has a name, knows it or at least could know it, and that it is socially acceptable to mention it to another person or even to a stranger).

[23] The same applies more or less to premises. However, premises are far more difficult to recognise than leading questions. They should therefore be avoided as far as possible in qualitative interviews. In addition, there is no need – in contrast to standardised interviews, for example – to set "decontextualised" frames. Rather, these frames or situation definitions are to be reconstructed in the research process and often form a central object of research.

Table 3.1 Specific questions, targeted text types and text type elements

Specific question form	Primary target text type	Primary targeted text type elements
Confrontations and polarizations	Justification	Justification
Summaries and conclusions	Validation and explication	Justification, explanation, confirmation/opposition
Leading questions and insinuations	Explication	
Hypothetical situations	Reflection	Justification

Form and Effect of Guideline Questions: Possible Errors in Stimulus Formulation

Within the framework of this presentation, it is not possible to go into all the difficulties, problems and conflicting goals of question and stimulus formulation, as these are legion. In principle, however, all formulations should be avoided which

1. unnecessarily limit openness and restrict the respondents' scope for articulation (Hopf 1978, p. 108), which is suspected for (uncontrolled) leading (suggestive) questions, yes/no questions and other closed questions as well as for normative-evaluative questions, among others;
2. may demotivate respondents in one way or another, which is intended to cause, among other things, linguistically complicated questions with difficult content, ambiguous questions, and questions that may trigger feelings of guilt or shame.

The "not to do" lists derived from this in guideline construction are generally useful and can provide problematic aspects of a question. Nevertheless, we would like to recall that the difference between formal question structure and communicative question function or communicative intention (and assumed effect), as already explained in Sect. 2.3 (see Table 3.2).

This will be briefly illustrated here using the example of yes/no questions. A formal yes/no question (and thus a formally highly closed question) can certainly be associated with a clearly more far-reaching communicative intention. Interview practice also shows that the expectation of receiving more detailed answers to a formally closed question is basically justified, although

(*continued*)

(continued)

Table 3.2 Formal and functional openness and closeness of questions

	Functional: open	Functional: closed
Formal: open	Please tell me about your childhood.	What is your name? Where do you live?
Formal: closed	Can you tell me about your childhood?	(Among other things, standardized questions with answer specifications)

of course not in all cases. (Conversely, formally open questions can also be closed in their performative logic).

Thus, for many questions it is not clear and often not easy to clarify whether and to what extent a question is open or closed in *its effect*. For this is often enough left to the interpretation and "willingness to cooperate" of the respondents. For example, the question "How are you?" can be answered with "good" or "bad", but can also be understood as an invitation to tell a story.[24] So before questions are sorted out as problematic or unusable on the basis of their wording, it should first be considered and, if possible, checked whether the formal question structure actually suggests a response behaviour corresponding to this structure.

3.3 Conducting Discursive Interviews as Guided Interviews

The numerous general problems and "traps" of conducting interviews have been described repeatedly and in detail, especially for guided interviews (cf. among others Helfferich 2009; Hopf 1978; Misoch 2015; Reinders 2005). The following account therefore concentrates on the difficulties that arise in the discursive interview, i.e. especially in the case of stronger interventions for the generation of justifications and other text types with a high proportion of derivations.

[24] This difference between the formal structure of a question or prompt has been part of the sociological body of knowledge since Garfinkel's (1967) breaching experiments.

3.3.1 Factors that May Influence the Interview Interaction

Section 2.2 already explained the basic approach of the discursive interview to the interactivity of data collection. It was made clear that, from the point of view of the discursive interview, this is not so much a problem as an opportunity to gain knowledge, which should be used constructively. Furthermore, the increased awareness of the importance of interview interaction that has become apparent in the more recent discussion of interview methods has led to a whole range of factors being considered problematic because they are thought to have an undesirable and uncontrollable influence. These factors can be broadly divided into two groups: These are, on the one hand, the social and individual characteristics of the participants and, on the other hand, characteristics of the interview interaction itself (see, among others, Holstein and Gubrium 2002; Misoch 2015, p. 199 ff.).

1. In particular, the list of social and individual characteristics of interviewers and respondents for which a detrimental effect is suspected is long. This includes all social characteristics of the interviewers, such as social gender, class, age and ethnicity, but also those that are less frequently used to characterise social differences, such as occupation, marital status, sexual orientation or place of residence. Interviewer effects, however, are also expected from personal characteristics of the interviewers. In addition to physical characteristics (e.g. attractiveness, disabilities, voice), these also include mental and psychological characteristics (e.g. speaking/questioning style, posture, attitudes, prejudices and prior knowledge of the interviewers).

   The characteristics of the interviewees that can have an impact on data gained in the interview are, in particular, their interview experience, their linguistic and narrative competence as well as their intrinsic motivation. Finally, particular importance is attached to the interviewer-interviewee ratio, i.e. the characteristics of the interviewers relative to those of the interviewees. This is considered problematic above all when corresponding differences reflect social power differentials (i.e. when women are interviewed by men, sick people by healthy people, migrants by non-migrants, etc.).
2. The interview situation itself can also affect the interviewing process in many ways. For example, factors are repeatedly mentioned that can trigger uncertainty in both interviewers and interviewees. These include the fairness dilemma (Hermanns 2000, p. 361), which is that, for all efforts, interviewees may be exploited or taken advantage of to some degree. Another problem discussed are the asymmetries between interviewers and respondents, which also exist inde-

pendently of social characteristics simply because the interviewers "define the situation". Finally, the permanent violation of the norms of reciprocity that apply in everyday conversations (interviewees should not ask questions themselves) and emotions that arise in the interview can also create uncertainty.

The relationship between the interviewer and the interviewee is also assumed to have a not inconsiderable influence on the process of data collection in the interview. Aspects such as sympathy/antipathy, closeness/distance or trust can have an effect here. However, the role patterns or role attributions by the interviewee (interviewer as "conscience", "helper", "therapist", etc.; cf. also Bogner and Menz 2001), which are often first to be defined (or repeatedly renegotiated) in the interview, are seen as central here. Finally, the location of the interview and the entire interview setting (e.g. the presence of third parties) are also considered to be factors that can influence the course of the interview in a potentially detrimental way (see Sect. 3.3.3).

The reasons why these and other factors influence the interview interaction and thus also the type of data obtained are quite plausible. However, so far there is no clear empirical evidence that the presumed effects actually occur.[25] There is also widespread helplessness with regard to possible solutions. Recommendations are usually limited to dealing with the problems consciously (e.g., about one's own possible effect as an interviewer), sensitively (e.g., in conducting the interview), and reflectively (especially in the interpretative analysis) (cf. Misoch 2015, p. 211).

Often these are possible influences that cannot be avoided or can only be avoided in part, but often enough are themselves a (partial) object of the research (and not disturbing variables). Sensitising oneself or the interviewers to the various relationship aspects in a research interview is certainly correct, but it is one of the usual requirements of qualitative interviewing.

If interviews are thoroughly planned – which, in addition to the construction of the interview guide, includes the training of the interviewers as well as pre- and post-interview support for the interviewees (including early clarification of the entire interview setting) – it should at least be possible to limit these influences.

[25] For standardized research, on the other hand, the occurrence of interview effects in particular has been sufficiently documented. It is doubtful, however, that these findings apply in the same or a similar way to qualitative interviewing (e.g. Misoch 2015, p. 210). This is because qualitative methods claim higher (ecological) validity and context sensitivity, which should actually prevent or at least significantly reduce such interviewer and situation effects.

Finally, the reverse can be said: if the gender of the interviewer, his/her knowledge advantage, the social distance between the participants, etc. have a detrimental influence on the interview interaction, to a considerable extent this will be due to deficient interview management.

Even more important for an analysis of collective representations with discursive interviews is that the assumption of problematic influences of social and other factors on data extraction in the interview is based on ideas of authenticity and truthfulness that are already contestable in themselves, but in any case are not relevant for discursive interviews (see Chap. 2). For even if the characteristics of interviewers can certainly have an effect on the *type of* derivations revealed in the interview, this does not change the fact *that* these derivations become evident and how they are structured. The influences resulting from the factors mentioned above are therefore also reduced to secondary questions of the distribution and perceived situational appropriateness of collective representations.

3.3.2 The Use of Guides in Discursive Interviews

Social factors and individual characteristics that can influence interview interaction thus do not pose any particular challenges to the Discursive Interview. The question of how the interactivity of the interview can best be used for eliciting individual derivations boils down to how to conduct the interview and to use the guide wisely.

First of all, an interview guide should be checked with regard to its applicability. This applies to each individual question or stimulus as well as to the partial and total structure of the guide. However, such a "pretest" interview is not a completed phase before the actual data collection. This is because the process of testing the questions and stimuli of a guide is never completely finished. It is rather a permanent readjusting the guide during the whole process of interviewing.[26]

On the one hand, the general manageability of the interview guide can be tested by applying it. For example, repetitions, redundancies, breaks (sudden changes of topic), but also the lack of communication elements such as transition, information and filter questions often only become apparent through the application test. In addition, such an initial "reality check" very quickly makes clear which questions or stimuli may not work despite the most careful construction.[27]

---

[26] This implies that interviews or at least parts of these interviews can be included in the data analysis even if changes were made to the guide after they were conducted. Nevertheless, a guideline should be checked particularly intensively at the beginning of the interviewing process so that it can be used largely unchanged in all interviews.

[27] For how to check guideline questions see Helfferich (2009, p. 182 ff.).

Possible weaknesses in formulation include (unintentionally) suggestive, over-demanding and moralising question formulations. In addition, unclear, ambiguous and normatively loaded terms can be identified at an early stage. Unexpected response reactions also reveal the multidimensionality of questions (in this case, the question is therefore ambiguous in content). It is particularly important to thoroughly examine pre-formulated hypothetical situations, confrontations and polarizations with regard to possibly problematic formulations, their content and, above all, their applicability. However, such a review does not only lead to the deletion and correction of questions; it can also result in suggestions for new questions or topics.

The most important basic rule for interview behaviour is that all questions and stimuli should always be used in a targeted and controlled manner. To put it pointedly: Everything is allowed as long as it serves the interest of knowledge and can be justified with it (and if no fundamental ethical norms are violated in the process). Furthermore, when dealing with the guideline in the interview, the danger of "guideline bureaucracy" already described by Hopf (1978) must be taken into account, which is particularly great in the Discursive Interview due to the active form of interviewing.[28] In addition, there are the problems already mentioned which can arise when using the specific instruments of the Discursive Interview, such as polarisation and confrontation in particular.

Perhaps the greatest difficulty for discursive interviews is therefore to make it clear to the interviewees that their perceptions and interpretations are in a sense always "correct" and precisely what the researchers are interested in. The interviewees should be "neutrally accepted" by the interviewers, so that above all they do not have to fear disapproval, but also cannot expect unquestioned acceptance of their narrations and justifications. This is obviously especially important when the reasons and derivations of the interviewees clearly deviate from supposedly "prevailing" or at least "widespread" and socially acceptable collective representations.

Especially in the case of rather "active" means of eliciting derivations, the interviewer should therefore never take on the role of an advocatus diaboli vis-à-vis the interviewees, always representing the opposite or deviant opinion, but also not that of a "confessor" or uncritical "claqueur". On the other hand, he/she should behave in a way that is recognizably neutral for the interviewees or – since this is only possible to a limited extent – consciously change these roles or make them flexible.

[28] In particular, the failure to miss a topic and the interruption of the interviewee should be mentioned here. The probability of both interviewer errors, which can only be reduced through appropriate sensitization, is likely to increase with the activity level of the interviewers.

Guide Design

Some questions also arise for the handling of the guide in the interview:

- The first concerns the technical form of *recording*: in addition to the classic forms (paper, index card), there are now above all various electronic options (smartphone, PC, tablet). Above all, however, there is also the possibility of not fixing a guide in writing at all, and only using it mnemonically.
- Secondly, it must be decided how *detailed* questions are to be fixed. These can be written down either completely or incompletely (e.g. only main questions). In the case of an incomplete question notation, the interviewers must be sure that the subordinate questions (less important, optional, follow-up questions to main questions) are nevertheless present to them in the interview situation.
- Finally, it is necessary to decide on the *accuracy of* the question notation. Thus, questions and stimuli can be formulated in detail (even with question variants) or noted down as keywords. In the second case, the exact question formulation then takes place (only) in the interview.

There are (too) many reasons for and against these and other elements of guide design and, above all, for the best combination of design elements. However, important decision criteria are certainly the respective research question, the number of guideline questions, the experience of the interviewers and possibly also the characteristics of the respondents.

For example, keyword-type questions have the advantage that they are formulated spontaneously in the interview and the danger of reading off and ticking off questions (guideline bureaucracy) is low. On the other hand, this form requires a lot of skill and self-confidence on the part of the interviewer, which is probably more often the case with experienced interviewers. Nevertheless, it can be useful to prepare question alternatives and to have all conceivable demand options "at hand". However, this can then lead to very long and unwieldy guides, which in turn make conducting the interview more difficult.

In short: Conflicts of objectives arise very quickly in the technical design of an interview guide. Each interviewer will have to find the best way for him/her. It is therefore reassuring that the technical design of a guide can of course also be adapted again and again in the course of the research. Thus it is relatively "natural" that interviewers become more and more confident in handling the guide in the course of the research and therefore increasingly condense it.

A "domination-free" communication atmosphere is certainly important for all interview forms, but for the discursive interview it is probably even more difficult to achieve than usual due to the "active" question forms provided. It is true that an open form of questioning and a correspondingly empathetic way of conducting the conversation are good (and indispensable) prerequisites for this. And even before the interview begins, it should be made clear, with reference to the special nature of interview communication and the neutrality of the interviewer, that the views of the interviewee are legitimate per se. Nevertheless, a "neutral acceptance" of the interviewee in the interview situation can be further supported by certain interview techniques. These include, above all, so-called "persil notes",[29] but leading questions (see above) can also be used in a similar way.

3.3.3 Other Aspects of the Interview Process

Two further aspects of conducting the interview are discussed here in conclusion, at least cursorily: the wider interview setting and the question of to what extent and in what form observations are made, noted and used during the interview.

The Design of the Interview Setting

The importance of the interview setting for the success of qualitative interviews should not be underestimated. There is no lack of recommendations for the (often technical) questions, which are full of details (cf. among others Gläser and Laudel 2009, p. 153 ff.; Reinders 2005, p. 147 ff.). Therefore, the presentation can be limited to a few points that are central to the Discursive Interview.

When designing the interview setting – which the researchers cannot determine unilaterally, but usually have a decisive influence on –above all the place and time of the interview must be determined; in addition, it has to be decided whether or how the interview will be recorded.

---

[29] "Persil notes" can be used to record views that the interviewees themselves may consider to be deviant, i.e. difficult to validate communicatively. In this case, the interviewer tries to give the interviewee the feeling, by means of assertions and careful insinuations, that even supposedly socially undesirable views can be expressed without further ado. In addition, the interviewer can also make the interviewee understand directly that he/she is quite familiar with supposedly deviant views and that they can be expressed. However, this requires a high degree of sensitivity on the part of the interviewer and corresponding information that can be taken from previous interviews, for example, or well-founded knowledge about the milieu, generation, etc. of the interviewee. Otherwise there is a danger of false premises and a methodologically and ethically problematic "fraternisation".

The success of qualitative interviews depends to a considerable extent on how comfortable the interviewees feel and as much they therefore feel free to express themselves fully and comprehensively. The best place for conducting discursive (and other) interviews is hence certainly the place where the respondents feel safe and have a high willingness to actively participate in the interview.

Basically, there are three alternatives for the interview location: here, there or somewhere else. "Here" refers to a room provided by the researcher. Such rooms (especially offices; possibly rooms equipped for interview recordings) have the advantage that the interview conditions can be optimally planned and controlled. "There" are the places where the interviewees like to spend a lot of time and which have been "appropriated" by them, so to speak, in their lifeworld. In most cases this is the home of the interviewee, but more rarely also, for example, the home of other people. These places are usually the ones where the respondents feel most comfortable and safe. Since they are also the easiest to reach for them, their own home is the interview location of choice for many respondents.

"Somewhere else" are all those places that are neither "here" nor "there", if you will: neutral places. These include cafés, park benches or rooms of the respondent's employer (e.g. teachers' rooms). However, such places have (almost) only disadvantages compared to the first two options. Above all, disturbances and the presence of third parties (see below) can hardly be controlled. They should therefore be avoided as far as possible, even though to respect the respondents' wish for the interview location will certainly increase their willingness to participate.

In addition to the central factor that respondents should be interviewed in a situation most comfortable for them, there are three other aspects of the interview location in particular that can influence the success of an interview because they can distract respondents: These are disturbing noises (pets, house and street noise, etc.), interruptions (smartphone, visitors, etc.), and the presence of third parties. All three sources of distraction should be excluded as far as possible.

Noise can be controlled to a large extent, but not always and best in the researchers' rooms. In the homes of the interviewees this is at least partly possible, but in other places it is usually almost impossible. A frequent, but at least partially remediable problem is also the underestimation of sources of disturbance, which are often far louder and more disturbing in the interview recording than during the interview itself, in which the participants can focus their attention accordingly. Finally, dealing with unexpectedly occurring noise can also prove difficult.

Interruptions are also best avoided in rooms provided (and possibly "prepared") by researchers. If interviews are conducted in respondents' homes, this can be addressed, at least in principle, while avoiding interruptions in public locations is more challenging. A particular and growing problem is interruptions and distractions caused by electronic media. In particular, turning off or even just not checking one's smartphone on a regular basis is apparently perceived as unreasonable by a growing segment of the population. This problem exists regardless of the location of the interview; however, the request to refrain from doing so seems to be most likely to succeed in rooms provided by the researchers.

The presence of third parties is perhaps the most serious of the three sources of problems mentioned: The (even temporary) presence of additional persons can change the interview situation so fundamentally (e.g. from an individual to a pair interview) that it exerts a considerable influence on what data is obtained. In most cases, moreover, neither the nature nor the extent of this influence can be determined ex post. Even if – depending on the research topic, the duration of a presence and the type of third person – the impact will vary, the presence of third persons should always be avoided. As with interruptions and noise, control is most likely in the researcher's rooms.

While noise and interruptions are similarly problematic for discursive interviews as for other interview forms, the presence of third parties poses less (and a different) problem here than in interviews intended to capture subjective meaning or authentic experience. For the derivations revealed are no less "true" if they are also addressed to a third person. A longer presence of a third person mainly leads to a new (triadic) constellation and presumably to different derivations being elicited than in the dyadic interviewer-respondent interaction. Nevertheless, these derivations are not "wrong". If, for example, the interviewee's partner is involved, it can be assumed that derivations become apparent that have been used and validated in the couples' private interactions.

All three "sources of problems" thus speak in favour of conducting interviews in the researcher's rooms. In determining the interview location, there is thus primarily a trade-off between conditions that are particularly easy to control and to protect against distractions (the researchers' rooms) on the one hand, and a location that is most comfortable for the interviewees (their home), going along with the hope of a high degree of openness and commitment on the other.

Far less complex than determining the interview location is the interview scheduling. Here, two points are primarily important: First, it must be ensured (or insisted upon) that the interviewees take sufficient time. They must therefore be realistically informed about the duration of the interview; ideally, they should not have a subsequent appointment. Above all, tight scheduling (lunch break, free period, etc.) should be avoided.

Secondly, respondents should be as alert as possible during the interview: Therefore, interviews should not be conducted at times that are inconvenient for the interviewees (early in the morning, late in the evening, etc.); likewise, they should not be conducted if they are ill or for other reasons not "fully capable of being interviewed" (e.g. alcoholic or emotionally stressed). In this case, as well as in the case of clear and early signs of fatigue during the interview, a new or additional interview appointment should be arranged.

Finally, a practical aspect that should not be underestimated is the form in which interviews are recorded.[30] First of all, it is important to have a technically good and reliable recording technique, and that the "human factor" (turning on the device, etc.) is also duly taken into account. Furthermore, it has to be decided whether the interviews will be recorded audio only or also video. The advantage of video recording is that non-verbal communication (facial expressions, gestures, posture) is recorded, at least in part. In contrast, audio-only recordings have the advantage of being less daunting and are more likely to be accepted by the interviewees.

The recording of non-verbal communication is only of minor importance for an analysis of collective representations. How derivations, when revealed in the interview, are accompanied by nonverbal forms of communication is not important for the reconstruction of derivations and collective representations. Nonverbal signals may contain additional cues (e.g., about the subjective meaningfulness of the derivations or about doubts about their validability); however, these do not refer to the derivations themselves, but "only" to their actual meaning for the interviewee. An audio recording is therefore sufficient for discursive interviews; on the other hand, there is nothing to be said against a (more information-rich) video recording, as long as this does not have a detrimental effect on the atmosphere of the conversation.

Context Information

In the literature on qualitative interview methodology, the collection of additional data is usually recommended (cf. among others Gläser and Laudel 2009; Helfferich 2009; Witzel 1985). In addition to key data of the life course and other "objective data", which are often collected in short questionnaires alongside the actual interviews, these include above all ethnographic contextual information of various kinds. Observational data refer to a considerable extent to the interviewee (including their appearance and behaviour), to the interview location and third persons, as well as to the interview communication and the interviewers. The corresponding observations are mainly recorded by the interviewers during and after the interview (or, if available, obtained from the video recordings). Other sources are preliminary talk and information from gatekeepers.

---

[30] The "whether" of a technical recording is not a question to be decided: transcripts, memory protocols, etc. cannot be used in the interpretative-reconstructive procedure of the discursive interview, which requires at least the exact wording for this (see Sect. 4.1).

Mostly such contextual information is used as an additional source for the interpretation of interviews. Such contextual information can also be important for the discursive interview and should therefore be recorded. However, in the Discursive Interview this is only the case for a very small range of this kind of information.

Thus, all verbal information that allows conclusions to be drawn about the interviewees' derivations should be recorded as accurately as possible. This can be particularly the case with "off the record" conversations (before and after the recording, during a preliminary talk, in e-mail contact), which can be very informative. Such conversations should therefore be noted in paraphrased form.

Furthermore, observations of the interview interaction can make the interviewees' statements more understandable and thus be important for the reconstruction of the individual derivations. The interview setting and interaction should therefore be recorded as precisely as possible. This includes the complete and accurate transcription of the interviews (cf. Sect. 4.1).

In contrast, all contextual information that goes beyond this and does not relate to the interview interaction is of no interest for discursive interviews. Collective representations and derivations are not to be found in the curtains, the furniture or the balcony plants of the interviewees – and they cannot be reconstructed any better, no matter how meticulously we record their appearance or consumer behaviour. (And despite the increasing body-sensitivity of qualitative research: for the reconstruction of collective representations, it is at least marginal whether and how these manifest themselves physically or in things).

All forms of arbitrary "ethnographic" information gathering with the aim of later use to contextualise the interviews should therefore be rejected, as should information that is typically gathered with short questionnaires for the same purpose. Not only do they not contain information relevant to the reconstruction of individual derivations and collective representations, but they may even encourage unfounded external contextualisation (cf. Sect. 4.3). Such information can therefore not be used meaningfully in the context of a reconstructive analysis of collective representations.

The basic rule for recording contextual information is therefore: it should be recorded as precisely and comprehensively as possible, but only to the extent that this can actually be justified with the goal of reconstructing individual derivations and collective representations.[31] Anything beyond this – at least from the perspective of the discursive interview – is voyeurism disguised as social research.

[31] Of course, it is often difficult to know in advance whether contextual information is relevant in this sense or not. (For example, even balcony plants might turn out to be important for the reconstruction of derivations). Therefore, the basic rule can be relativized in this way: Contextual information should always be collected in case of doubt, but should be used as sparingly as possible, namely only if it is certain that no external data are arbitrarily used for contextualization (cf. Sect. 4.3).

The Discursive Interview as a Face to Face Interview

Until now, it has been "taken for granted" that the Discursive Interview is a (co-present) face to face interview. In fact, the Discursive Interview is designed for face to face interactions. The specific question types and frequent follow-up questions assume the possibility of immediate clarifications and "repair actions", which "classically" exist in face to face situations. For this, para- and non-verbal signals are particularly important, which presuppose co-presence or visual contact between interviewer and interviewee. For this reason, discursive interviews – like most qualitative interviews – cannot be conducted as telephone interviews or in written form.

In view of the new ways of communication that have emerged with the Internet (Web 2.0), however, the question arises anew as to whether discursive interviews can also be conducted in forms other than face to face. However, little is known so far about how online media can be used for qualitative interviewing (cf. Früh 2000; James and Busher 2009; Mann and Stewart 2000, p. 195 ff.). This applies even more to whether and how online interviews can be used to record individual derivations.

In principle, however, there is much to suggest that forms of Internet-based communication can also be used for qualitative research. For example, the high level of anonymity and the great flexibility of online media are often seen as advantages for qualitative interviews. The fact that many online media require "low-threshold" written communication (especially emails and chats) also opens up new and interesting options for qualitative research (cf. Schiek and Ullrich 2016; Ullrich and Schiek 2014).

For discursive interviews, the lack of visual contact is a particular problem. This is only possible, at least to a large extent, in the case of video conferences or chats, even if communication with para- and non-verbal signals is inhibited during a virtual presence. All (primarily) written forms of communication (especially e-mails), on the other hand, make it considerably more difficult to conduct discursive interviews: On the one hand, the danger of misunderstandings that cannot be cleared would be very high. On the other hand, it is to be feared that the pressure to reflect in written communication reduces the chance of revealing individual derivations.

For the time being, it can therefore be assumed that individual derivations can best be elicited in face-to-face situations. Especially written and asynchronous forms of communication lead to less spontaneous but more reflected and abstract justifications and will therefore tend to contain less collective representations.

3.3.4 Aspects of Research Ethics

Questions of research ethics have always accompanied social research. This is also true, and perhaps even especially so, for qualitative research, for which a concern with aspects of research ethics can be already be found in (and in the discussion of) classic studies such as "Die Arbeitlosen von Marienthal" (Jahoda et al. 2001) and "Street Corner Society" (Whyte 1996).

Recently, however, questions of research ethics have come much more to the fore (they even seem to completely dominate some "postmodern" discourse). The reasons for the increasing importance or problematisation of research ethics issues in social research are primarily to be found in external stimuli such as publicly scandalised practices (for example in medical research) and an increasing awareness of the dangers associated with the mass storage of sensitive data. This has led to growing pressure from the public and politics (including via research funding institutions such as the German Research Foundation [DFG]) and ultimately to many professional societies such as the German Sociological Association (DGS) formulating a discipline-specific code of ethics (1993).[32]

At least in the social sciences,[33] research ethics differs from professional ethics, which, among other things, regulate interactions between colleagues, in that it is essentially concerned with shaping the research relationship between researchers and "researched persons" and, in turn, particularly with the handling of information and the protection of data.[34] The central aim is to avoid disadvantages and risks for research participants, which should be fully anticipated by the researchers as early as possible (von Unger 2014, p. 19). Research ethics requirements often conflict with research interests, but do not necessarily have to do so.

---

[32] How compliance with research ethics standards can be monitored or guaranteed has so far only been partially resolved. The instruments that already exist include, above all, declarations of commitment by researchers (e.g. in the case of research funding by the DFG), information and ombudsman offices at professional associations and ethics committees.

[33] Research ethics problems vary greatly according to discipline and type of method. They certainly do not arise as urgently in the social sciences as they do in medicine, for example, and they are quite different for standardized methods than for qualitative ones. The different qualitative approaches also raise different issues here. For example, it can be assumed that research ethics requirements lead to more conflicts with research interests in ethnography and participatory research than in interview methods.

[34] According to Hopf (2000, p. 589 f.), in the social sciences the term research ethics covers "all those ethical principles and rules in which it is determined in a more or less binding and more or less consensual way in which the relations between researchers on the one hand and the persons involved in social scientific investigations on the other hand are to be arranged."

With regard to the discursive interview (but also to qualitative social research as a whole), three aspects of research ethics seem to be of particular importance[35]: questions of confidentiality and anonymity, the voluntary nature of participation, and (beyond that) harm avoidance:

Perhaps the most difficult problem and one of the most delicate tasks in qualitative research concerns issues of *anonymisation*. While general confidentiality is comparatively easy to ensure and maintain (e.g. that information is not passed on to relatives or employers), anonymisation of participants presents particular challenges in qualitative research.

Thus, one should always be aware that, due to the small number of cases and the type of sampling, true anonymity is not possible in qualitative research. It is therefore all the more surprising how often precisely this is promised (e.g. in internet templates). However, only anonymisation is possible, i.e. a procedure that offers participants as much protection from identification as possible (e.g. by deleting real names). But even then, the participants can often be easily recognized by persons to whom they are known, due to the comprehensive and dense data material. Of course, this is especially true if research participants deanonymise themselves, e.g. by telling their acquaintances about their participation in the research (a problem that is all too rarely pointed out). Rules of anonymisation should therefore be very strictly adhered to. At the same time, however, no untenable guarantees should be given: It is more correct here to inform the participants comprehensively and in good time about the probability, the reasons and about possible consequences of identification.

Voluntariness is a basic prerequisite for interviews and all other procedures that require the direct participation of the research subjects, because without a corresponding willingness, research interviews cannot be conducted in a meaningful way.[36] The usual solution here is "informed consent". This consists of informing the respondents as comprehensively as possible about the research background, the interview type and the interviewers, as well as about the role, significance and tasks of the respondents and possible consequences of participation. If the interviewees

---

[35] For further aspects of research ethics see, among others, Hopf (2000) and von Unger (2014).

[36] This also applies to groups of persons for whom the consent of third parties is required (including minors and persons in custody). This consent is only an additional hurdle and does not change the fact that the interviewees must want to cooperate. Voluntariness is moreover fundamentally more precarious with these groups of persons and requires a special degree of consideration on the part of the researchers.

want to participate, they are no longer only superficially, but "well informed" willing to do so and the criterion of voluntariness can be considered fulfilled.

"Informed consent" must also be seen as a process. It must be renewed again and again, e.g. if respondents become aware of the possible consequences of their statements during the interview. A declaration of consent given in advance, even if written, would be worthless in such a situation; rather, "informed consent" must then be obtained again.

This often leads to a trade-off with methodological requirements. For example, consent that can be revoked at any time certainly restricts spontaneity and possibly makes (especially discursive) interviewers proceed more cautiously than would be desirable. However, a the informing interviewees comprehensively can above all have an (undesirably) strong structuring effect. For if the respondents know what the researchers are interested in, they will possibly orient their answers accordingly. Very often, therefore, there is a methodological interest in keeping respondents at least somewhat in the dark about the "actual" research goals.[37]

Even informed consent does not release researchers from their responsibility to *avoid harm* (to participants, but also to third parties). This applies not only to the case that respondents underestimate the consequences of their participation and could harm themselves as a result, but also to the further course of the research, which cannot be anticipated in all details in advance.

For interviews, this means on the one hand that the process of data collection (i.e. the interview conversation) itself must be considered as a possible source of damage. Thus, every interview (and the Discursive Interview possibly more than others) causes the interviewees to reflect on themselves, their biography and their current life situation. In addition, the Discursive Interview, through its specific question forms, is designed to challenge common sense prejudices and theories, which can also lead to destroy them for the interviewee. Positively, every qualitative interview is a small learning process for the interviewee and a chance for self-reflection. It is evident that this can also lead to stress or even personal crises. Emotional reactions in the interview are certainly the most important indicator for this; however, their "calming down" alone may not prevent medium-term damage.

The principle of damage prevention must also be followed in later research phases. This applies above all to publications and other forms of knowledge transfer, as well as to data archiving, where questions of anonymisation again arise in particular (How can a high degree of anonymisation be achieved in publications?

[37] A possible solution here is an ex post declaration of consent. In case of refusal, however, this leads to a complete loss of data.

How can interview recordings and transcripts be archived with low risk?). In principle, however, emotional damage and other consequences that are problematic for interviewees are still possible (e.g. if an interviewee reads the interpretation of her statements). In publications, too, care should therefore be taken to anticipate and avoid harm to respondents as much as possible.

When conducting discursive interviews, therefore, several aspects of research ethics are affected all at once. These require researchers to deal responsibly with "their" participants. The first step in this process is always a corresponding self-sensitisation of the researchers. However, we should not forget that participants are generally sensible people who can and want to decide for themselves about the risks they take by participating in the research process.

Analyzing Discursive Interviews

4

With discursive interviews the "core business" of data analysis is the interpretation of the interview transcripts with the aim of reconstructing the "structures of meaning" (in this case, the collective representations). However, the process of data analysis is much more comprehensive and can be roughly divided into three phases:

Before the interpretation and reconstruction of meaning structures, the interview material must first be prepared for such analyses. This primarily involves transcribing the interviews, which is explained in Sect. 4.1. Coding the transcripts is also part of the preliminary work. Since coding in the understanding used here is not independent of interpreting, it will also be explained in this context (Sect. 4.2). The essential goal of data analysis is, of course, the reconstruction of collective representations . Often, however, researchers will not be satisfied with this and will, in further analysis steps, attempt to condense these reconstructions into types and, if possible, to develop a typology of collective representations (Sect. 4.4). However, before this last step of a reconstructive data analysis based on Discursive Interview is explained, the question of how different types of context can be included in the reconstruction of collective representations will be briefly discussed (Sect. 4.3).

4.1 Transcription

If for the Discursive Interview the contrasting interpretation of the interviews is the main task of the analysis process, the most important requirement for this is that they are available in written form. For only on the basis of a written version systematic comparisons of text passages can be meaningfully carried out. The transcription (writing down) of the audio- or video-recorded interviews is therefore the most

C. G. Ullrich, *The Discursive Interview*,
https://doi.org/10.1007/978-3-658-38477-7_4

important preliminary work for the reconstruction of collective representations by means of a contrasting interpretation.

The basic aim of transcriptions is to "translate" oral into written speech as accurately and loss-free as possible. The most important of these "typically oral" features are non-verbal (gestures, facial expressions, posture), para-verbal (including laughter, sighing) and prosodic elements (such as intonation, volume, pauses in speech) of the spoken language. In addition, sounds (e.g. clapping, knocking) and special features of the situation (e.g. interruptions, presence of third parties, sources of interference) are usually taken into account in the transcription.

In addition to the form and accuracy of a transmission of non-linguistic features of oral communication, the required notation style must be determined. Here, a decision has to be made between a line-by-line and a score notation, in which it is presented on a time axis who speaks at which time, so that, for example, simultaneous speaking and interrupting can already be recognized in the form of the representation.

There are several alternatives for the accuracy with which the sound of oral language is transcribed. These range from standard orthography to various literary and phonetic transcriptions that more or less accurately reproduce specific features of the spoken language (such as "musta been" instead of "must have been"), to pure phonetic transcriptions (e.g. the International Phonetic Alphabet).

There are a number of different rules and techniques for transcription, often developed in linguistic contexts, as well as specialized transcription software (for both, see Dittmar 2009).

When deciding on a transcription system, it should be borne in mind that as accuracy increases, so does the effort required for transcription. Since this happens in an almost exponential manner, before transcribing the interviews, it should be carefully considered which features of oral communication that are typically missing in written language need to be transcribed. This is mainly determined by the research question, but also by the specific method applied. Especially if linguistic aspects play a role here, the demand for a thorough transcription will be rather high.[1]

[1] In view of the effort involved, one should be cautious with excessive forms of transcription that are in no way required for the interpretation, but instead are used to demonstrate some kind of (pseudo-)scientificality.

Elements of a Transcription

The general aim of transcribing oral interviews into written interview transcripts is to make the special features of oral communication recognisable in the written text as well. In addition to the question of whether to transcribe line by line or in score notation, the following aspects in particular must be taken into account:

- phonetic authenticity: accuracy with which the oral language is reproduced (including dialect, specific speech codes, individual peculiarities of pronunciation and speech errors)
- non-verbal communication: gestures, facial expressions and body posture (only possible with video recordings)
- paralinguistic features: including laughter, sighing, moaning, crying, coughing, short laughs
- prosodic features: esp. stresses, (increasing/decreasing) loudness, stuttering, (increasing/decreasing) speech tempo, pauses in speech (measurement of pause lengths), anacolutes
- Sounds accompanying conversations such as clapping or drumming (e.g. on tables)
- interpretative comments on linguistic expressions (usually by the interviewer) such as "ironic", "anxious" or "angry".
- Specifics of the speech situation: including interruptions, presence of third parties, external sources of interference (pets, doorbells, street noise, etc.)

Whether these features of oral language are depends on what is required for the research. If the decision is positive, transcription rules must be defined for the representation of these features; together they form the transcription system.

For all kinds of interviewing it is true: The more accurate transcription is the better one, because less information is lost. On the other hand, elaborate transcripts will contain much information that is of little importance for the reconstruction of collective representations. For example, a continuous score notation is not necessary. For the identification of individual derivations, however, it is important that the transcription shows exactly to which stimulus (part) the interviewee is reacting (e.g. if interviewers are interrupted by the interviewee before the question is finished). Also, an elaborate rendering of sounds is of little value for research interested in collective representations. A simple literary transcription should therefore suffice as a rule. (Nonetheless, a notation in standard orthography is to be rejected as "distortion").

A complete and exact verbatim transcription of the interview communication, on the other hand, is indispensable for the analysis of discursive interviews. Non-verbal (especially facial expressions and gestures) and paralinguistic features (e.g. laughter, groans) can be largely dispensed with, because or insofar as they do not contain any information that is important for the reconstruction of derivations and collective representations. It is more important to fix prosodic aspects as precisely as possible (e.g. pauses, sentence breaks and stresses), which (like the exact time of speech) clarify the elicitation process of the derivations. Thus, prosodic features such as anacolutes or stresses can provide clues as to how collective representations are used in justifications.

All in all, the demands on interview transcription in the Discursive Interview are moderate, especially in comparison to linguistic studies and interview procedures that aim at the authenticity of experiences and actions. If, however, doubts about the transcription quality should arise, there is still the possibility of "listening" to the recorded conversations with pinpoint accuracy (and, if necessary, re-transcribing them).

4.2 The Reconstruction of Collective Representations from Interview Transcripts

4.2.1 Contrasting Interpretation

The comparative or contrasting interpretation of interview transcripts is the central step in the reconstruction of collective representations in the Discursive Interview and will be explained in this chapter. To this end, this section first deals with general questions, before Sect. 4.2.2 sets out the concrete methodological procedure of reconstructing collective representations using discursive interviews.

Understanding – Interpretation – Reconstruction

To interpret can be defined as a purposeful, non-intuitive process of understanding meaning (or meaning-transporting objects). Accordingly, the result of this process can be called *interpretation. Understanding* (Verstehen), in turn, can be defined as the immediate-intuitive grasp of a meaning (possibly a meaning context) of a human artifact (e.g. a statement). Intuitive understanding is the "everyday" form of comprehending subjective meaning, since immediate (at least approximate) comprehension is a basic prerequisite for interactional participation. Interpreting is a "normal" competence and activity of socialized interaction participants: All too often, meanings (especially the meaning of other people's actions) are not self-evident and force a corresponding process of reflection.[2]

Interpretation in the field of qualitative research is essentially about a methodically controlled understanding of others that should be intersubjectively comprehensible. This means above all that the interpretation must follow rules and be plausible (also for potential recipients). It is therefore important, among other things, that the rules of interpretation are disclosed and that the interpretations are verifiable (comprehensible) for others.[3] If more than the comprehension of a subjective meaning is intended, one is dealing with *reconstructions* of structures of meaning (in the case of the discursive interview, therefore, of socially shared knowledge structures).[4]

More complex systems of rules for interpreting texts are often called *hermeneutics*. Depending on their professional origin and theoretical background, these can differ significantly. The procedures provided for in hermeneutics are usually understood as "experience approved interpretative rules" that are intended to stimulate and support the work of interpretation. They cannot and should not be used like recipe books or instructions for use. For even the most exact adherence to interpretive rules and the most slavish adherence to all rules do not by themselves lead to interpretations for which the repeated and "circular"[5] engagement with the text always remains decisive.

Many interpretation rules and techniques are based predominantly on positive experiences in research. Mostly, they are not epistemologically based. If we know, or believe, *that* the compliance with certain interpretation techniques leads to viable interpretations, then this is based on that our experiences have shown or even proven this. But then we usually do not know *why* this is so.

---

[2] The definitions given here are only to be understood as working definitions. The processes of "understanding" and "interpreting" have received a great deal of attention in the social sciences in general and in qualitative research in particular, and have been repeatedly redefined in different ways (see, among others, Danner 2006; Graumann et al. 1991; Hitzler 1993; Ricoeur 1972; Soeffner 1989). Likewise, different interpretive systems or hermeneutics have been developed (cf. among others Hitzler and Honer 1997; Jung and Müller-Dohm 1993; Soeffner 1979).

[3] This is the essential difference to, for example, literary interpretations. In the case of literary interpretations, the creative contribution of the interpreter is of greater importance, which is why very different interpretations of the same text are possible and deviations or new interpretations are desirable.

[4] This seems to me to be the rather common use of the terms "reconstruction" and "reconstructive", which are usually not further explicated, but nevertheless often used to distinguish reconstructive from other qualitative methods.

[5] This circularity finds expression in the metaphor of the "hermeneutic circle" (cf. Danner 2006, p. 60 ff.).

Basic Forms of Interpretation

In interpreting, three basic strategies can be distinguished, which are more or less emphasised in most hermeneutics:

1. The first is based on the distinction between *manifest* and *latent meaning:* in addition to what is "obviously" said or written (manifest meaning), a further (latent) level of meaning is assumed to underlie the manifest meaning that the sender is aware of. Especially for structuralist, psychoanalytic and, in a broader sense, "ideology-critical" approaches, the manifest-latent difference is of central theoretical and empirical importance. Corresponding research interests aim to reconstruct the structures of meaning that are unconscious to the actors involved themselves, as well as their "generative" or leading effects. This is usually attempted through very detailed interpretation of selected and elaborately transcribed data material. The starting point are usually case analyses, which are carried out sequentially (i.e. in the sequential order of text).
2. Possibly the most widespread interpretation strategy is the interpretation of meaning with the help of *contextual knowledge*: The meaning of a statement or text passage is inferred from the context in which it is found. What is defined as context and which contexts are finally "consulted" can be very different and range from the textual context (e.g. the immediately preceding utterance) to the speech situation (e.g. an interview) to cultural and historical contexts. Contextualising interpretive strategies therefore differ in terms of the contexts considered important or legitimate, as well as the timing and form in which contexts are used interpretively (see Sect. 4.3).
3. The third basic form of interpreting is based on the distinction between individual and social meaning and an interest in reconstructing social structures of meaning. The interpretation strategy for determining social meaning (e.g. collective representations) consists of a *systematic comparison (contrasting)* of different text passages. Through repeated and varied comparisons (including cross-comparisons), similarities and differences and, as a result, socially overarching patterns are made clear.

These three basic forms of interpretation are not mutually exclusive, but are linked in various ways. Depending on the objective and methodological basis, however, different emphases are set. The reconstruction of collective representations on the basis of discursive interviews also draws on all three forms of interpretation. As already pointed out in the first Sect. 1.2, contrasting is something like an "interpretative meta-strategy" for the discursive interview.

Contrasting Interpretation

The basic idea of contrasting interpretations is to arrive at new insights through systematic and varied comparisons of data material. In contrasting procedures, they form the (eponymous) *basic strategy of interpretation*, without the other two basic forms therefore being meaningless. Comparisons are thus carried out in order *to interpret* and not following an interpretation. This distinguishes contrastive from many other interpretive procedures, which provide for comparisons, but without these being central to the interpretation itself.

Comparing ≠ Comparing

Comparing or contrasting is undoubtedly one of the most common procedures in qualitative social research. There are two reasons why the comparison of text passages as a means of interpretation is so emphasised here: Firstly, it is often not clear enough what is being compared and why; secondly, it is nevertheless clear that comparisons in qualitative social research are often carried out for other purposes and therefore at other times and at other levels than in the Discursive Interview.

These differences also exist beyond the decisions on criteria and strategies of concrete comparisons (see below) and go back to fundamental methodological considerations and conceptual determinations:

1. With regard to purposes, at least three further "comparative goals" can be identified in addition to the comparisons for interpretation provided for in the Discursive Interview. Especially following sequence-analytical interpretations, but also in other areas in which the reconstruction of single cases is in the foreground, comparisons serve to increase the scope of already completed case reconstructions or to control or interpretatively secure the case analyses. Comparisons are thus used here for the generalization or validation (cf. also Chap. 5) of research results. The third comparative objective is to use comparisons of codes and categories for the development of (empirically based) theories, and is prominent above all in grounded theory methodology (Strauss and Corbin 1990).
2. Closely related to these objectives is the level (or time of research) at which comparisons are made. In the case of case comparisons for generalisation or validation, these always take place after the case reconstructions have been completed and thus also after the interpretative-reconstructive analyses have been completed. Comparisons as a strategy for theory building, on the other hand, are carried out primarily at the level of codes and categories previously formed by the researchers on the basis of text passage analyses. In

(*continued*)

(continued)

Table 4.1 Levels and purposes of comparisons in qualitative social research

	Discursive interview (analysis of collective representations)	Sequence analysis procedure	Grounded theory methodology
Level/point of comparison	Text (identically coded passages from interviews)	(completed) case reconstructions	Codes and categories
Aim of the comparison	Interpretation (and reconstruction)	Generalization and validation (case reconstructions)	Empirically based theory building

both cases, therefore, it is not the original text material that is compared, but intermediate products (codes/categories) or end products (case reconstructions) that have already been generated in the research process.

Comparisons in the Discursive Interview thus differ fundamentally from these forms of contrasting both in terms of the objective (in the Discursive Interview: interpretation of interview transcripts) and the level of comparison (in the Discursive Interview: comparison of interview passages) (Table 4.1).

Contrasting procedures thus differ categorically from comparisons after a completed interpretation, because only here are comparisons a ("the") *means of interpretation*. The level of comparison is therefore never the "case" in the sense of a document (in the case of interviews: an interview transcript). Comparisons are rather carried out "below" this level and insofar across the document units.

The "method of comparison" (Mill 1843; Durkheim 1984) is one of the oldest methods in the social sciences and is still widely used today in various disciplines. Traditionally, comparisons at the macro level and the use of statistical methods are dominant here (e.g. in "comparative political science"). As a qualitative procedure for evaluating or interpreting texts, contrasting (interpretive) procedures have been described only rarely or incompletely (cf. Kelle and Kluge 1999). This is quite surprising in view of the (at least repeatedly claimed) paramount importance of contrasting and case comparisons, especially since these had already been described and explicitly propagated in some classic works of qualitative social research.[6]

[6]This is of course especially true for grounded theory, for which Glaser and Strauss (1967, p. 105 ff.) regarded the "constant comparative method" as fundamental for the development of theories close to the subject. Therefore, it is quite correct to see comparison as a core heuristic principle of grounded. On the other hand, it should not be overlooked that comparisons as a methodological procedure have clearly moved into the background in the (self-)representations of Grounded Theory Methodology and, in some cases, are no longer addressed at all (see, among others, Breuer et al. 2018; Charmaz 2014; Dey 1999; Strauss 1991; Strübing 2008).

In general, however, at least four typical steps of contrasting procedures can be distinguished for the field of qualitative methods (cf. also Kelle and Kluge 1999, p. 38 ff.): (1) First of all – attention has already been drawn to this in Sect. 3.1 – the course for the desired contrasting should already be set during sampling. This can be done, as shown, on by preliminary theoretical considerations ("selective sampling") or (better) in the form of "theoretical sampling", in which the sampling process is guided by the need for new and further possibilities for comparison. (2) The second step consists in coding the text material. The codes used can be based on theoretical considerations or developed "inductively" from the text material (and thus in the research process itself). (3) On the basis of the material thus coded as completely as possible, identically coded text passages (i.e. text passages that are considered comparable on the basis of the code assigned to them) are then compared for interpretation, including across the original document units. (4) Finally, on a second, more abstract level, the codes and categories (summaries of several codes) newly obtained or modified from these comparisons can be compared again.

Comparison Criteria and Strategies

If data are to be analysed in a contrasting manner, two fundamental questions must be clarified: that of the comparison criteria and that of the comparison strategy.

Comparison criteria or categories are those by which the coded text passages are compared. The code assigned to the passages forms the tertium comparationis. Thus, text segments are compared which have been defined as comparable and got the same code. The coding of the text material is thus on the one hand a central for the preparation of interpretative comparisons, and on the other hand already a first step in interpretation (for types of coding see also Sect. 4.2.2).

Do Contrasting Procedures Tear Apart the "Natural Context of Meaning"?

A frequent criticism of contrastive interpretation accuses them of destroying natural contexts of meaning; and indeed text segments are detached from their textual context for interpretation. The aim of such a "cut and paste" approach (before electronic text processing, transcripts were actually cut up and pasted together from other points of view) is essentially to reassemble parts of the text.

The accusation of disregarding a "natural" context of meaning (actually: of the genesis of meaning) is nevertheless based on a threefold misunderstanding:

(*continued*)

(continued)

1. First of all, it is questionable whether or in what sense a context of meaning exists in guided interviews. If transcripts of guided interviews are the source material, a context of meaning can only be assumed in a very limited form. In guided interviews, for example, new stimuli are constantly being set (and thus new meanings are constantly being generated); the answers are therefore not in a "natural" context (one that develops independently of researcher intervention), but, if at all, in a controlled and reactive one. Guided interviews such as the discursive interview are also a form of conversation, but a particular one, aimed at contrasting options. Guided questions are therefore often not a follow-up communication resulting from the conversation, because they are also intended to enable different (and later contrastable) aspects to be addressed in the interview.
2. Nevertheless, there is also a more or less strong connection of meaning between the individual interview answers, which interviewees often also make explicit by referring directly or indirectly to earlier answers. Thus, the questions asked first will have an effect on aspects addressed later in the interview. However, this can certainly be taken into account in contrasting procedures: On the one hand, by explicitly focusing on relationships between codes during coding, so that even "intratextual-intertextual" references can be captured and interpreted in a comprehensive form (and not only in their sequential order). On the other hand, a broad form of coding makes it possible to include at least the immediate interaction context (in interviews usually question and answer) "untouched" in the contrasting interpretation.[7]
3. More fundamentally and crucially, however, the claim that contrasting interpretations disrupt the ("natural") context of meaning conceals a subjectivizing or individualizing understanding of meaning. Therefore, this critique may apply when the goal is to capture subjective worlds of meaning (as in psychoanalytic and social work case analyses), but not when social contexts of meaning are to be reconstructed. For social meaning, according to the central premise here (cf. Sect. 2.1), is not hidden in a

---

[7] It should be self-evident that when coding answers, these are not separated from the questions or stimuli, because responses cannot be understood without the associated question. Nevertheless, "question-free" coding and citation of answers seems to be very common.

(continued)

single text (interview transcript), but only manifests itself in parallel in different texts: only their common basic structure, freed from individual idiosyncrasies, can be considered social (shared). Contrasting interpretations therefore aim precisely at uncovering this "social essence" in the individual applications of meaning (here: derivations).

In short, there is no question of "tearing apart" the context of meaning in contrasting interpretations, because there is no such textual context. On the contrary, this social meaning must first be "distilled" and reconstructed from the different texts through comparative interpretations. The corresponding accusation is thus based on a solipsistic concept of meaning that fails to recognize the social character of knowledge.

In addition to the comparison criteria, the *comparison strategy* must also be defined, which does not preclude (often this will be desirable) carrying out several comparisons with different comparison strategies. Thus, on the one hand, rather similar (minimal comparison) or rather different text passages (maximal comparison) can be compared. A minimal comparison is particularly suitable for condensing initial interpretative ideas, whereas a maximal comparison is used primarily to highlight differences and to delineate the entire subject matter. In contrasting interpretations, minimization and maximization strategies are, however, only a question of the interpretatively more favorable starting point for comparisons; for ultimately all text segments are included in the contrasting process.[8]

4.2.2 Contrasting as Interpretative Strategy of the Discursive Interview

The interpretation of the interview transcripts is an integral part of the discursive interview and forms a functional unit with the sampling and the interview procedure, which is oriented towards the reconstruction of collective representations. Fundamental to this is the view, as explained in Sect. 1.2, that collective

[8] In addition, text passages can be compared both within the same text source (e.g. an interview transcript) or across text sources (comparison of identically coded text passages from different text sources). For contrasting interpretations aiming at social meaning, cross-text comparisons are especially important.

representations can best be captured through the systematic comparison of individual derivations.

Contrast is made by comparing text segments below the level of the interview transcripts (and therefore across them). Thus, it is not "cases" in the sense of interviewees or interviews (interview transcripts) that are compared, but shorter text units. The criterion for comparison is the common theme or reference problem, or the derivations related to it. Often this is just because text segments are answers to the same guideline question.[9] This systematic comparison of text segments is used as a means of text interpretation and is the crucial instrument for reconstructing collective representations from individual derivations.

Contrasting understood in this way fulfils three important and intertwined functions in the context of analyzing discursive interviews:

1. First of all, contrasting is fundamental for the *discovery and reconstruction of collective representations*. For only by comparing and "superimposing" can the collectivity of representations be recognized. A collective representations is thus reconstructed by comparing all the individual derivations of a reference problem[10] in terms of similarities and differences. This is done in the form of a synopsis,[11] in which common basic structures are uncovered through overlapping processes of comparison. As soon as patterns in the interpretations, explanations and justifications become repeatedly visible (and across the interviewees), there is an "initial suspicion" of a collective representations.
2. In most cases, synoptic comparison will reveal more than one collective representations. The reconstruction of the distinct collective representations related to the same reference problem must therefore be imagined as a parallel process. Through the crystallization of a collective representation and the visualization

[9] For an analysis of collective representations with discursive interviews, it is not the bearer of a representation but the collective representations that constitute the "case" to be reconstructed. In this sense, one could also speak of contrasting of cases. However, since this usually refers to comparisons of already analysed research units, the form of interpretation of the Discursive Interview is referred to here only as contrasting (without case).

[10] The "reference problems" to which the individual derivations refer are for the most part already apparent from the topics set by the guide. In addition, the respondents can of course also refer to new phenomena not anticipated in the guide or redefine reference problems. In addition to the collective representations and derivations, therefore, the problems of reference (as perceived by the interviewees) must also be reconstructed to some degree.

[11] On synoptic comparison, cf. also Kelle and Kluge (1999, p. 56). Synopsis is the systematic comparison of texts in order to be able to recognize differences more easily (e.g. synopses of legal texts) or to uncover a common "generative" structure (classically: synoptic gospels of the "New Testament"). The latter is also the primary goal of contrastive analysis here.

of differences, the entire *set of collective representations* of a reference problem finally takes more and more shape. This is also the first step to generating types (cf. Sect. 4.4).

3. Finally, the parallel reconstruction of all collective representations of a reference problem also secures those of each single collective representations through repeated comparisons. It thereby increases the stability and *credibility of* the single interpretations

The aim of the contrasting interpretation is thus the reconstruction of the different collective representations and their interpretative validation as well as a first step towards the development of a typology of collective representations.

In iterative processes of comparison, the collective representations are therefore reconstructed interpretatively from the individual derivations. However, the reconstruction of the individual derivations is more than an intermediate step in the reconstruction of collective representations. Rather, it is a parallel process in which both derivations and collective representations are reconstructed through separating the individual-particular and the social-general. One could thus say that the reconstruction of individual derivations is an (unavoidable) by-product of the reconstruction of collective representations.

Coding, Recoding, Relationing and Grouping

Of the steps or levels of contrasting described above (Sect. 4.2.1), only the second and third are contrasting *interpretations* as understood here.[12] *Coding or indexing* texts is therefore the central procedure for preparing and carrying out contrasting interpretations.

Code and Coding

"Code" and "coding" are differently used and often irritating and fuzzy terms. One of the reasons why these terms are irritating is that they are also used in many other contexts (intelligence services, biology, linguistics, etc.). Even more important – and possibly easier to confuse – is that coding is also

(continued)

[12] Sampling strategies such as "theoretical sampling", on the other hand, are an previous process in which systematic variations can only be made at the level of the units of inquiry (respondents). As with comparisons at the level of codes or code groups (categories), comparisons here are not a means of interpretation, but a separate (non-interpretative) step in the analysis.

(continued)

used in standardised research, where the assignment of numerical values to characteristic values (e.g. 1 for "yes" and 0 for "no") is referred to as coding.

Coding as a technique of qualitative social research was made known primarily by Strauss (1991) and Strauss and Corbin (1990) within the framework of the Grounded Theory Methodology he developed. However, what is referred to as coding in each case in different qualitative studies is often not very clear. This is complicated by the fact that, in addition to code and coding, seemingly similar terms such as category/categorization or concept are used, and it is often not clear whether and in what ways coding differs.

In qualitative social research, however, (at least) two basic procedures are referred to as coding: One is the assignment of codes to text passages presented here in *preparation for interpretations*. In addition, coding can denote *a method of interpretation*. Thus, multi-stage coding according to Grounded Theory Methodology is to be understood as a procedure for obtaining empirically founded theories, in which coding itself describes the process of interpretation.

"Indexing" is a term proposed by Kelle and Kluge (1999), which expresses more clearly than coding what is also involved here, namely the pure assignment of codes to text passages. If possible, all text passages should be assigned at least one code, but they can also be coded several times. For the comparative interpretation, all passages coded in the same way are then used.

The codes assigned to the text passages should always be obtained "inductively" from the text material. Nevertheless, it is likely that many codes can also be traced back to guideline questions, and are to a certain extent pre-determined by them. Which coding style (e.g. keywords, paraphrases, numbers) is used depends not only on personal preferences but also on the research question. In any case, the variety of coding styles used in research is surprisingly large (cf. Saldaña 2009).

Coding or indexing the text material in the Discursive Interview does not represent an actual interpretation process and thus differs from (some) other coding procedures. It only serves to prepare synoptic comparisons. But even if it is not intended, codings are always already initial interpretations (and should therefore be conscious as such). This is obvious when the codes are developed "inductively" from the textual material, for every "designation" of a passage and even the determination of the textual unit being coded are already interpretative acts. But it is true even if the interpretation is only in the form of an assignment of text to a pre-existing code. (For most codes, both are true anyway: even if they were first developed from the textual material, other passages of text are then assigned to the already existing codes.)

Before the "actual", contrasting interpretation, the coding thus already forces a first interpretative step. This is not unproblematic, especially because "clumsy" coding may obscure the view for comparative options. Unfortunately, there are many possibilities for problematic codes; but three forms in particular seem to be common:

On the one hand, these are codes that are (too) theoretical and "deductive" (e.g. technical terms, theoretical concepts, but also everyday terms) and can thus limit the spectrum of possible interpretations at a very early stage. Such theory-guided codes are not only the consequence of "deductive" coding (which should be avoided anyway), but can also occur when the codes are at least apparently obtained from the material. Thus, a completely theory-free coding is not possible, since this would presuppose a theory-free language. One should therefore always be aware of this and try to avoid "uninformed" and "unreflected" theorizing as much as possible.

The other two coding pitfalls are that codes are either too broad (comprehensive) or too narrow. Very narrow codes lead to too few comparisons being possible (in extreme cases, a code is assigned once), and in the opposite case of very broad, comprehensive codes, the amount of text can quickly become so large and so heterogeneous that comparisons are hardly possible.

Moreover, whether codes "work" in the sense of a contrasting interpretation cannot be determined in advance. At the latest, however, the attempt to carry out a contrasting interpretation on the basis of the codes assigned will very quickly reveal the inadequacies of individual codes. Repeated recoding will therefore almost always be unavoidable.

The codes developed slowly in this way can then be combined into larger groups and related to other codes (even outside these code groups).[13] This elaborate and repeated process of code systematization leads to a comprehensive network structure of all codes, which will also be partly hierarchical (main and sub-codes).

The repeated grouping and relating of codes as well as the necessity to recode text passages and reformulate codes require much more effort than the original coding. However, recoding, grouping and relating codes are indispensable sub-

---

[13] This corresponds roughly to the procedure of "axial coding" of the Grounded Theory Methodology (Strauss 1991, pp. 56 ff., 101 ff.). Unlike the "coding paradigm", however, the focus here is primarily on the thematic proximity and hierarchizations of codes. Causal and other relationships between codes, on the other hand, are of rather little importance for the reconstruction of collective representations.

steps of interpretation because they force an intensive interpretation of the data material. All this results in the determination of the comparative criteria necessary for the contrasting interpretation. The fact that this often requires a reorganization of the code hierarchy and the code network is a logical consequence of these iterative processes.[14]

How necessary and useful this procedure is for the reconstruction of collective representations cannot be judged in general terms. To a considerable extent, this is determined by the first (original) coding. For example, relationing of codes is more likely when using narrow codes than when using broad ones, where possible relations may already be contained "in" the (single) code. Similarly, narrow codes are more likely to have possibilities or a need to group them. Finally, recoding will be necessary especially in the case of "theoretically biased" and inconsistent (initial) coding.

The "Actual" Interpretation

Coding (indexing) the interview material is thus the most important step in the preparation of a contrasting text interpretation, but at the same time it is always a first sub-step of the interpretation, because the naming and organization of codes already require interpretative efforts and structure the subsequent contrasting. The last interpretative step consists in the elaboration of a set of collective representations, which – for reasons to be explained – results in real type formation and is presented in connection with questions of type formation (Sect. 4.4). In between lies the "actual" (or real) interpretative process, i.e. the interpretative reconstruction of individual derivations and collective representations on the basis of the developed codes. This interpretative core process is ultimately a creative act that cannot be learned or achieved with a to-do list. Qualitative social science interpretations are also always dependent on intuition and therefore remain subjective to a certain degree.

In the qualitative literature, however, many interpretation rules have been proposed and tested that can at least facilitate this "creative" part (cf. in more detail Mayring 1983; Oevermann et al. 1979; Strauss and Corbin 1990). Five of these suggestions or "golden rules of interpretation" are particularly recommendable with regard to a reconstructive development of collective representations:

---

[14] Nevertheless, there may also be singular codes that do not fit into any group and have no relationships to other codes. These may, but must not indicate a coding deficit.

1. The first and most important is that of a *"interpretation-stimulating"* mindset. This consists mainly of two aspects: The first is the open attitude of the researcher. This includes in particular that research findings are not hastily subsumed under favourite concepts and theories. Only those who approach the interview material in a theoretically unbiased way can really discover something new in it. But openness also means being curious and attentive. Again, if one is only moderately interested in one's own research results, it will be difficult to use the material creatively. The second aspect of an attitude helpful for interpretation concerns the "attitude to work": convincing reconstructions will only be achieved by those who are prepared to "submerge deeply" into their interview material (interpreting is above all a lot of work).[15]
2. *Paraphrasing* text passages can prove to be extremely useful. Similar to coding, it can help to find access to the material and to first ideas for interpretation, because paraphrasing is already a "small interpretation". Above all, however, it has the great advantage that can be done even when the external circumstances (time, motivation, etc.) are rather unfavourable (paraphrasing can actually always be done).
3. For the reconstruction of collective representations, it is also important to *proceed* close to the text and *with textual accuracy*. Thus, the answers of the interviewees should always be taken literally. Linguistic characteristics (Why was the statement formulated in this way?) should therefore be taken into account interpretatively. To this end, it can also be helpful to compare the genuine with alternative formulation options.
4. The basic form of empirical text comparison can be supplemented by *thought-experimental variations*. In addition to actual differences, differences constructed theoretically and, in particular, an exploration of the space of possibilities can help to clarify the particularities of the reconstructed knowledge structures.
5. Finally – and this has already been pointed out in Sect. 2.3 – the *interview interaction* should be included throughout: It is particularly important to treat the interviewees' answers as such (i.e. as reactions to the interviewer's questions) and not as statements without context. Such an embedding in the interview context is always necessary for the interpretation. Being aware of the interactive context of a statement can also facilitate interpretation. Furthermore, possible effects of earlier questions and topics in the interview as well as of the entire situational context (interview setting) should also be included interpretatively.

[15] Such basic attitudes are of course difficult to plan, but should always be considered. If one is occupied with too many other things, one should perhaps refrain from interpreting.

Software for Qualitative Data Analysis

The use of software for qualitative research[16] is not one of the strategies that can help with the "actual" interpretation. On the contrary, software can (mis) lead to neglecting the original material.

Nevertheless, the use of a software program is recommended, especially for extensive text material, as long as it is limited to its core function. This is the so-called coding and retrieval function, i.e. the management and retrieval of text passages on the basis of codes that have previously been assigned to the text passages by the researchers. The clear advantage of software programs is that they relieve researchers of tedious tasks (e.g. searching for text passages). In addition, software programs make it easier to organize data, increase flexibility (e.g., by allowing text passages to be referred to again and again quickly and simultaneously in different contexts), and reduce the risk of data being lost or "overlooked".

Software programs thus offer assistance in handling the text material, that should not be underestimated. However, qualitative software programs definitely cannot interpret.

4.3 Excursus: Contextualizing in the Analysis of Discursive Interviews

The importance of context for the interpretation of qualitative data has already been emphasized many times and contextualization has been characterized as a fundamental form of interpretation. For the reconstruction of collective representations, too, the inclusion of different contexts is an important building block of the contrasting analysis. However, this does not apply to all forms of context. This chapter will therefore clarify when and which form of contextualisation is methodologically appropriate for the reconstruction of collective representations.

[16] There are many software programs to support qualitative data analyses, which differ only slightly (cf. Kelle 2002, among others).

4.3.1 On the Meaning and Determination of Context(s)

Context, contextuality and contextualization are central concepts in qualitative social research. Malinowski (1923), for example, had already emphasized the importance of context, and in particular situational context, for the understanding of verbal utterances. The inclusion or appropriate consideration of context as a prerequisite for understanding has always been a central postulate of all interpretive approaches (cf. among others Hoffmann-Riem 1980). Conversely, the ignoring of contexts or the deliberate attempt to minimize contextual influences has been seen as one, if not the decisive problem of standardized survey research (cf. among others Cicourel 1974; Garfinkel 1967; Wilson 1982).

In German methodological literature, too, the importance of context for interpretation can be considered "canonized" (cf. among others Hoffmann-Riem 1980; Steinke 1999, p. 28 ff.). Lamnek (1988, p. 25), for instance, emphasizes that the "meaning of an action or a verbal expression becomes intelligible only by recourse to the (symbolic or social) context of its appearance". And with regard to interviews, Kohli (1978, p. 13) claims: "The analysis of the situational context of the interview is crucial for determining to which everyday situations the respondent's utterances can be generalized." Similarly, Bohnsack (1991, p. 21) states, "In the single interview, I can only adequately understand the specific utterance in the overall context of a narrative or longer account."

Despite this generally high appreciation of context, further explications are rather rare. As self-evident as the importance of context is considered to be, as unquestionable it seems to be what the context of a text is, how this context can be found and how it can be included in the process of interpretation. In this context, the three rather randomly selected quotations already make it clear that, with the same appreciation, quite different contexts can be meant (here, after all, already four, namely a symbolic, a social, a situational and a narrative context).

A general treatment of the question of how and which context is to be taken into account in interpretations may also not initially seem sensible. The reason for this is simple: as different as the data and approaches to reality are that are pursued with qualitative methods, as different may be the contexts that are used for interpretation. This is also true for qualitative interviews, which can not only vary greatly in form, but are also used on the basis of divergent methodological considerations.

Accordingly, the different qualitative methods and methodologies would be the place to define and specify context. But even here, no systematic discussion of questions of context is found,[17] especially not in interview methodology.

Context Types

What (a) context is is quite ambiguous. Accordingly, many different types of context are defined and classified differently. However, the various context models converge on some points. On this basis, three types or levels of context can be distinguished, at least roughly:

1. The first is the textual *context* (also co-text). This is primarily the text before a passage to be interpreted, whereby a distinction must be made here between the "immediately before" and all "further before" text passages. The textual context also includes the text (immediately) after the passage to be interpreted as well as the text as a whole (i.e. the interview transcript in the case of interviews). Finally, other texts can also be defined as "intertextual" co-text (e.g. when interviewees refer to such texts).
2. The second type of context is the *situational-generative context* of a text. This includes, on the one hand, the communicative genre, i.e. whether a text is an interview, a magazine article or a private telephone call. More important here, however, are the specifics of the text's genesis, the importance of which has already been pointed out in Sect. 2.2. In the case of interviews, this is above all the interview setting (time and place of the interview, way in which the interview is conducted, etc.).
3. As the third and broadest type of context, *external contexts* are also distinguished. These can be very different and include not only the individual and social characteristics of the "text producers" (in the case of interviews, among other things, the age, gender, social status, etc. of the interviewees) but also the broader social, political or even historical context. The determination of external contexts considered relevant thus has hardly any "upper" limit.

[17] The two main exceptions here are conversation analysis and the rather "context-skeptical" stance of objective hermeneutics (cf. Oevermann 1997; Wernet 2000, p. 21 ff.), i.e. two methodological directions that are not exactly known for a particular affinity with interview methods. In international and linguistic discourses, too, questions of defining and differentiating context(s) have long been neglected. Clark and Carlson therefore also speak of context as a "conceptual garbage can" (1981, p. 314), a characterization that is still almost proverbial today.

Even if a method-independent determination of contexts does not seem possible, general *requirements for a methodologically sound determination of contexts* can be formulated. Thus, for the determination of relevant contexts, answers must be found to at least four sub-questions (cf. e.g. Franck 1996):

1. First of all, and this is by no means trivial, it must be clarified what is to be considered a *text*, for which the context is then determined.[18] "Text" is, on the one hand, to be determined substantively-materially, especially if the concept of text also includes (textualized) actions or objects (e.g. images). But even in the case of texts in the narrower sense, a more precise definition of what the text is supposed to be may be necessary. For example, in the case of interviews, it needs to be clarified whether non-verbal and para-verbal elements are to be understood as part of the text, as context, or perhaps as both at the same time. On the other hand, it has to be clarified whether or when text units have "own" contexts or whether context is always only that of a text entity.
2. Secondly, it must be clarified what is to be considered context for a chosen text. That a context becomes a context only through its relation to the text seems evident. The question of *determining relevant contexts* is incomparably more difficult and controversial. Central to this is the question of what (and how much) contextual information is necessary for understanding a text. Accordingly, several proposals for differentiating different types of context have been made (cf. among others Auer 1996, p. 16 ff; Goodwin and Duranti 1992, p. 6 ff; Harris 1988; Reisigl and Wodak 2009, p. 93).
3. Once text and context have been determined, the *relationship between text and context* must be clarified. The main issue here is the question of analytical primacy: is contextual information used sporadically and selectively to interpret single passages in the text, or is the text understood as a "typical" expression of the context that is actually of interest? Thus, although the text defines the context, the latter precedes the text in time and causally.[19]
4. Finally, it needs to be clarified whether the context of a text must itself also be contextualized. This raises the question of the "*contextualization of contexts*", which is important for the methodological safeguarding of interpretations, and also how to deal with the danger of an infinite regress.

[18] For a general definition of "text" see, among others, Brinker (2005) and Fix (2008); on the concept of text in qualitative social research, see especially Soeffner (1979) and the contributions in Garz (1994).

[19] Thus, Heritage (1984, p. 242) can be agreed with when he states that "the significance of any speakers communicative action is doubly contextual in being both context-shaped and context-renewing". But this also means that contexts are only partly – and probably to a much lesser extent – generated or modulated in the situation of text production.

4.3.2 Implications for an Interpretative Interview Methodol ogy

The fact that contexts have to be defined and justified method-specifically, but also touch on fundamental questions of qualitative social research, is also shown by the increasing preoccupation with questions of context since the 1990s (cf. esp. Duranti and Goodwin 1992; Miller and Dingwall 1997), which has led to sometimes very controversial debates (cf. in summary Blommaert 2001; Mey 2001). At the core of this was the question of who decides what is included as context: Is it (also) the researchers who have to ensure appropriate contextualization, or is it, and exclusively, the "text producers" involved who consciously or unconsciously provide contextualization cues.[20] If one follows the arguments of this controversy, the dilemma of contextualization is that one must either limit oneself to the participant contexts or set contexts "deductively" (theory-guided) as researcher contexts.

For interpretative approaches and approaches interested in obtaining interview data, however, both approaches are not feasible. As important as the contextualization cues of participants are for interpretative analyses, context cannot be equated with participant context. This may make sense and be necessary as long as the research interest is directed towards the "how" of the mechanisms of "conversational management". However, as soon as socially shared meanings is the central research goal, this cannot do without the inclusion of further contexts, even if the interviewees themselves do not give any corresponding contextualization cues.

Equating context with the participant context is also particularly problematic for the research interview as a form of communication. This is because, unlike in recorded conversations, for example, interviewers take an active part in communication and are therefore involved in both the interactive constitution of meaning and the constitution of context in the interview – and all the more so in the way they structure the conversation through their contributions, as in discursive interviews. The contextualizations in the interview are also influenced to a considerable extent by the "institutional situation" of the interview (the interviewees know what a research interview is and that they are participating in one).

Both the active participation of the interviewers and the "institutional situation" of the interview make it impossible to treat qualitative interviews like (recorded, non-reactive) everyday communication. Thus, a solution to the problem of context

[20] The first position has been advocated mainly by discourse analysts, the second is a widely shared and often emphasized view among conversation analysts. For example, Wolff (1986, p. 67) states, "Context is regarded in a conversation analytic perspective as a performance of the conversation or its participants. It is not the context of the conversation, but the context in the conversation that is of conversation analytic interest."

determination has to be found which, on the one hand, does not limit contexts to the participant or production context from the outset, but, on the other hand, still determines them empirically, i.e. from interpretative analysis, and avoids "deduced" contexts. However, there is still no satisfactory solution in sight for the question of how contexts are to be determined "correctly" and, in particular, also within the framework of the analysis of collective representations.[21]

For the time being, this means for the methodological foundation of contextualizations in reconstructive analyses of qualitative interviews:

- What counts as context cannot be determined independently of the research question and methodological position. (There are no "objective" contexts). The minimum requirement is therefore to disclose how contexts are defined in interpretations.
- Contexts cannot be determined externally within the framework of interpretative and reconstructive strategies, but must in any case be worked out from the empirical material or from the text in order to avoid arbitrariness.
- Moreover, an empirical determination of the relevant contexts always requires decisions on the part of the interpreters. Through this alone, but also through participation in the situation of creation, contexts are always co-"constructed" by the researchers.
- The empirical anchoring in the text is not to be equated with the participant contexts, nor is there a need to determine contexts for the single text. Especially for comparative interpretation strategies, this increases the scope for a nevertheless empirically based context determination.

For the reconstruction of collective representations on the basis of discursive interviews, it can finally be stated that the narrower textual and situational contexts are to be taken into account in principle, external contexts, on the other hand, only in the case of corresponding contextualization cues by the interviewees.[22] In other

[21] However, proposals such as those of Stenvoll and Svensson (2011), which are based on a procedural strategy, are promising here. They propose a stage model of textual contextualization that does not limit it to participant contexts. Specifically, they distinguish three types of contextualization cues, namely explicit, implicit and indirect. This form of contextualisation seems to be more advanced because it does not confine the interpreters to a passive role, but above all because contexts can also be obtained here in contrasting textual analyses. At the same time, this makes it clear that contextualizations that go beyond participant and situation contexts do not have to be derived "deductively".

[22] A more comprehensive consideration of external contexts, on the other hand, may become necessary if the reconstruction of collective representations is carried out in conjunction with a discourse analysis (Keller 2014). In this case, the extended need for contextualization results from the discourse-analytical requirements.

words: The risk of using contexts that cannot be legitimized methodologically because they are arbitrarily set by the researcher in the reconstruction of collective representations increases with the distance from the original text and from the immediate interactive situation in which it was created and should therefore, if at all, be done with restraint and reflection.

4.4 Type Construction

Terms such as type, type construction, typology or even ideal type are used a lot in the social sciences, often with very different intentions and very often with a rather vague definition. Qualitative social research is no exception. Therefore, a few preliminary conceptual clarifications are needed here first. These will not be able to remove the many ambiguities of the extremely "tricky" discussion on type terms and type construction; their sole purpose is to deal with the question of type construction in the context of an analysis of collective representations by means of discursive interviews without causing misunderstandings too quickly.

4.4.1 Real Types and Ideal Types

In general, two basic forms of types can be distinguished. The first form will be referred to here as the real type(s). The second is the famous ideal type.

Real types are attempts to structure empirical diversity by means of conceptual constructions. It is characteristic of real types that they capture only the "full", empirically verifiable forms as they can be reconstructed from the data material. The prefix "real" expresses here that empirical phenomena can be meaningfully designated and summarized with this term.[23]

Real types are usually achieved by grouping (or "clustering") "cases" that are similar in terms of a number of characteristics. What is defined as a case in each case depends on the research question and whether a contrasting approach is taken or via individual case analyses. (In the Discursive Interview, it is the derivations applied to a reference problem). However, such a grouping of cases alone is not sufficient. Only when it is possible to uncover or make clear a common, fundamental structure for these groups of cases can one speak of the formation of one or more real types.

[23] The term real type was first used by Eucken (1947, pp. 68 ff.; 418 ff.) in deliberate juxtaposition to Weber's ideal type, but is much less common today than the ideal type term. The distinction made here between real (empirical) and ideal (theoretical) types largely follows Eucken's distinction.

Real types, even if they emerge from contrasting analyses, can be single types. Very often, however, several more or less clearly distinguishable real types are built. In this case, one can speak of a set or set of real types. Such a set of real types should represent as well as possible the field under investigation, or at least the research results from which it was developed, and thereby improve the understanding of that field. If, on the other hand, real types are not comprehensible or not well distinguishable from others, the real type set has failed. There is no claim beyond this, especially an applicability to other fields. Above all, this distinguishes real types from ideal types.

As is well known, the concept of the *ideal type* goes back to Max Weber (1995), who has repeatedly given it new and slightly different versions. For Weber, ideal types are above all a heuristic instrument intended to support research and theory-building by, as it were, sharpening the view of reality.

Ideal types are "ideal" because they are theoretically "clear" – and so internally consistent that real cases never, or only extremely improbably, correspond exactly to an ideal type. Ideal types therefore never consist of arbitrarily selected and unrelated features, even if these are empirically present, but contain explicit and explainable relationships between their components, thus contain at least the nucleus of a hypothesis. Ideal types are thus clearly more presuppositional than the construction of real types (Table 4.2).

The importance and use of ideal types are justified by their heuristic value: "In order to see through the real causal connections, we construct unreal ones" (Weber 1995, p. 287). Ideal types are therefore also not a reflection of reality, but are intended to "lend unambiguous means of expression to its representation" (Weber 1995, p. 190). The formation of ideal types is thus justified by the fact that they contribute to a better understanding of a state of affairs than would be possible without such a formation of types.

According to Weber (1995, p. 191), an ideal type "is obtained by the one-sided enhancement of one or a few points of view and by combining a plethora of diffuse

Table 4.2 Real and ideal types

	Real types	Ideal types
Destination	Structuring and explanation of the research results	Heuristic tools; theory building and guidance of (future) research
Procedure	Empirical reconstruction (esp. by contrasting methods)	Theoretical construction (if possible on the basis of empirical research, possibly real types)
Single types	Empirically reconstructed structure (based on one or more cases)	Through "unrealization": theoretically substantial concept
Several types	Structuring of the investigated subject area (real type set)	General conceptual heuristics ([ideal] typology)

and discrete, here more, there less, in places not at all, individual phenomena, which conform to those unilaterally emphasized points of view, to form a unified body of thought. In its conceptual purity, this image of thought cannot be found empirically anywhere in reality; it is a utopia, and for historical work the task arises of determining in each individual case how close or how far reality stands to that ideal image (…)."

Already in Weber (1995), two basic types of ideal types can be distinguished (cf. on this also Mommsen 1974, p. 221 ff.; Pfister 1928, p. 170 ff.). On the one hand, these are historical or single types. Single ideal types are mainly used in historical (diachronic) research interests (e.g. epochal concepts such as capitalism) and usually serve to explain complex processes of social change and historically enduring structures. Examples from the field of collective representations analysis are the historical analyses of the genesis and development of far-reaching collective representations, such as those by Honegger (1978) on the early modern collective representation "witch" and by Schütze (1986) on the collective representation "mother love".

If, on the other hand, the research interest is comparative, one will usually not be satisfied with individual types. The aim is then rather to develop a *typology*,[24] i.e. to structure a subject area as completely and theoretically as possible.[25] One can speak of a complete typology if it covers the whole field, i.e. if all possible or necessary ideal types have been formed. In this case, there is no single phenomenon that cannot be described and explained with the typological conceptual tools. Whether or not a typology is complete can ultimately only be shown by the test of experience (its heuristic proof). In principle, however, it must be assumed that typologies have a rather provisional character.

Just as the development of a single ideal type is more presuppositional than that of real types, so too the demands made on a typology far exceed those made on sets of real types. It consists above all in the fact that the conceptual tools developed must prove their value for theory building or empirical research "independent of time and place". Well-known examples of typologies that have been successful in this sense are Weber's types of rule (1985) and Esping-Andersen's (1990) typology of welfare states.

---

[24] The term typology is restricted here to ideal types, so the prefix "ideal" is not necessary here.

[25] Typologies may consist of logically interdependent ideal types in which the single ideal types are meaningfully related to each other, as is often the case especially in antagonistic-complementary typologies (such as dichotomous class models) or in stage models. However, interdependent ideal types will be the exception, because such relations between ideal types will often not be achievable.

Whereas sets of real types are intended to structure a concrete subject area and represent it conceptually, typologies aim at a heuristic set of instruments that is valuable for the analytical penetration of social realms of reality even beyond and independent of the context of their origin. Many theoretical concepts are therefore ideal types in this sense. Real types and sets of real types can be a preliminary stage of the formation of ideal types, i.e. ideal types can be developed out of existing real types.

Typologies vs. Classifications

Part of the confusion about typologies certainly results from the fact that classifications are repeatedly passed off as typologies (cf. Barton and Lazarsfeld 1979, p. 59 ff.; Hempel 1976; Kelle and Kluge 1999, p. 77 ff.; Kuckartz 2010, among others). Thus, classifications formed from two variables are often referred to as typologies and are then usually also presented in cross-tabular form (in the minimum case of two variables with two characteristics as a four-field table).

Classifications, however, follow a completely different logic than typologies. They aim to completely cover a range of possible combinations of characteristics. Therefore, the requirements of uniqueness, exclusivity and completeness also apply to classifications (cf. Friedrichs 1973, p. 88 f.). According to this, each object of a feature space must be able to be assigned exactly one feature combination; it must therefore fit exactly into (only) one "class". The possibility of empty classes is a logical consequence of this structuring logic. An example of a classification are multidimensional models of social stratification (which take into account e.g. income, education and social prestige): In such a model, each individual must be able to be assigned to exactly one stratum (and not to several at the same time). The aim of a classification is to illustrate the possible combinations of characteristics in a field and to "sort" the units of study accordingly, often in preparation for a quantifying study (e.g. on stratification). Classifications always presuppose that the relevant characteristics and their expressions are already known. They are obtained deductively and can be presented in the form of lists and tables.

Classifications can therefore also be "meaningless", because in principle any feature can be combined with any other. They do not convey any substantial statements about social reality. The usefulness of classifications results solely from their usefulness for descriptive and quantifying analyses. The decisive factor here is that the relevant characteristics are already known. Types and typologies thus differ from classes and classifications primarily on the basis of two properties: First, they should be "meaningful", i.e. show specific relationships between their elements; second, the features relevant in this context are determined in the process of analysis, i.e. reconstructed from the data material.

4.4.2 Type Construction in Qualitative Social Research

In qualitative social research, the formation of types is very widespread and is usually seen as a central strategy for generalising research results and for quality assurance (cf. among others Flick 1995, p. 255 f.; Przyborski and Wohlrab-Sahr 2014, p. 376 ff.; Rosenthal 2008, p. 74 ff.) or simply for structuring one's own research results (cf. Kluge 1999, p. 43 ff.).[26] Often, however, the "crowning" of qualitative research with (ideal) types seems so self-evident that the purpose of type construction is not even explained in detail.[27] The same has been repeatedly stated for the procedures of type formation. Whether it is part of the "peculiarity of any type construction (…) that its implementation is extraordinarily difficult to explain and rationalize" (Wagner 2008, 213), or whether other reasons are responsible for this, cannot be judged here.

Even though this is rarely made explicit, qualitative-empirical types are predominantly real types. This is also obvious insofar as real types are a means of structuring and explaining a subject area and should therefore also be suitable for structuring research results.[28] Moreover, real types are the logically closer or first form of type construction as a result of qualitative social research.

[26] Becker's concept of "constructed types" (Becker 1959, p. 155 ff.) and the considerations of Barton and Lazarsfeld (1979) were also influential for the reception of type concepts in qualitative social research. One of the earliest qualitative works to explicitly perform ideal type construction was by Gerhardt (1984, 1986). On the history of the reception and application of type concepts in social research, see especially Kluge (1999).

[27] Kluge (1999) and Kelle and Kluge (1999), for example, give only extremely vague indications of why types are formed in qualitative social research, namely "in order to grasp complex social realities and contexts of meaning and to be able to understand and explain them as far as possible" (Kelle and Kluge 1999, p. 75). The usefulness of type concepts for different forms of scientific analysis (i.e. also for the interpretation and classification of qualitative research results) is hardly disputed; however, it does not justify why types (of all things) should be aimed at as a result of qualitative-empirical research. Kluge (1999, p. 51 ff.) also presents in great detail different uses of the concept of type in qualitative social research. However, such conceptual explications are obviously fruitless and only show that the concept of type (with different prefixes) is used almost arbitrarily in qualitative social research.

[28] Przyborski and Wohlrab-Sahr (2014) also come to this assessment, according to whom "the construction of ideal types should not be understood primarily as a result of research, but above all as a method of research" (2014, p. 379). However, it does not logically follow from the correct statement that ideal types are a heuristic instrument that they cannot be developed empirically from research results.

The different interpretation and analysis strategies thereby also lead to different types, which corresponds to Weber's distinction between single types and typologies. Case- or sequence-analytical interpretation procedures usually aim at case or single types, whereby the already completed interpretation of the "cases" (e.g. of life stories) forms the starting point. Such forms of (real) typification serve in particular to generalize general (cross-case) structures (cf. inter alia Gerhardt 1986; Rosenthal 2008, p. 74 ff.). Contrasting procedures, on the other hand, usually aim at several real types related to each other. The formation of types is thus here a part or the conclusion of the contrasting interpretation procedure.

However, typologies based on contrasting procedures or procedures that see themselves as contrasting typologies have repeatedly caused irritation. This is mainly due to the aforementioned conflation of typologies and classifications (e.g. in Kelle and Kluge 1999, p. 77 ff.; Kluge 1999, p. 31 ff.; Kuckartz 2010) and the associated use of terms (such as cluster; feature space), which many people tend to associate with standardised methods. However, the frequent, almost apodictic and not further substantiated insistence on individual case analyses as a starting point for (then case-) contrasting type formations (e.g. Kelle and Kluge 1999, p. 75) is also irritating – after all, this is diametrically opposed to the logic of contrasting interpretations.

Furthermore, the admission, if not explicit demand, of "deductively" obtained comparison criteria (e.g. Kelle and Kluge 1999, p. 83 ff.) and thus also of the basic construction characteristics of types, which is very offensively advocated especially in the field of the documentary method (cf. e.g. Bohnsack 1989, 1992; Nentwig-Gesemann 2001; Nohl 2013), must be regarded as particularly problematic. Such an approach, an external pre-determination of the basic criteria for type construction, is incompatible with the "inductive" and open approach of qualitative social research and is excluded here as a possible approach to type construction.

Real types and especially a convincing reconstruction of a set of real types can be considered a successful outcome of qualitative studies. Nevertheless, real types can and are often seen and used as a starting point for an ideal type formation.[29] This is because the formation of ideal types or an (ideal) typology is often sought as a (more ambitious) goal of qualitative analyses.

[29] In principle, classifications can also be derived from real type sets, whereby real type sets can be regarded as incomplete classifications. Insofar as the real types were obtained through empirical research, however, this would be disappointing as a result, to say the least. This is because classifications can be formed further with less effort from simple combinations of features.

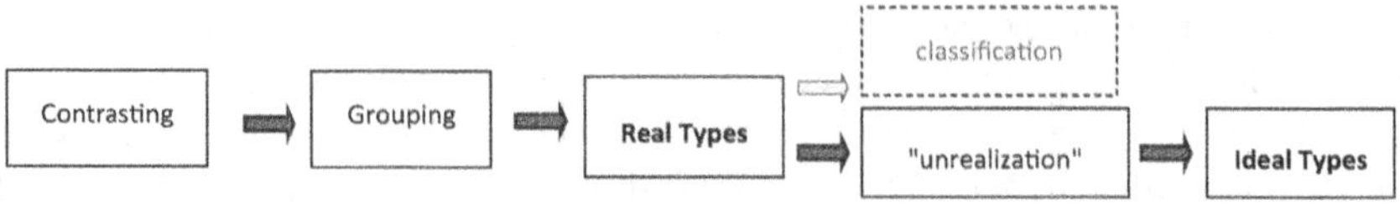

Fig. 4.1 Real types and ideal types as results of a qualitative research process

Weber's explanations of the formation of ideal types (see above), however, first of all make two aspects clear that should at least make one skeptical with regard to the aim of striving for ideal types as the end point of qualitative-empirical research: On the one hand, ideal types need by no means or even only be obtained empirically, but can also be developed theoretically, as Weber himself predominantly did; and if they are based on empirical research, quantitative – Esping-Andersen (1990) typology of welfare states is a prominent example of this – as well as qualitative data can form the basis for them.

More importantly, however, the formation of ideal types is primarily an analytical process (of "cleaning up", "thinking ahead", etc.). Ideal types cannot be obtained directly interpretatively (albeit on the basis of empirical real types), but are the result of an additional, theory-constructing analysis of the material at hand, be it real types or other "intermediate products". An interpretative reconstruction of meaning structures can therefore never lead directly to the formation of ideal types; this always requires an additional analytical step (cf. Fig. 4.1).

Nevertheless – this will be explained for the analysis of collective representations – the formation of ideal types as a (moreover appealing and demanding) objective of qualitative research should not be abandoned lightly. There is little doubt that a successful formation of ideal types or even a typology can very significantly increase the value of a qualitative study because they elevate its significance above the narrower research context. On the other hand, however, there seems to be just as much doubt that qualitative research results are very often not suitable for the formation of ideal types.

4.4.3 Type Construction in Collective Representations Analysis (on the Basis of Discursive Interviews)

The application of typologizing procedures in the course of an analysis of collective representations is to a certain extent unusual. For collective representations are already "typified forms of knowledge" that depict more or less complex contexts in a simplified way and, precisely because of this, make the users of the collective representations capable of acting. Collective representations are therefore themselves already types. This peculiarity, however, is only a semantic problem and has no influence on the procedure for the formation of real types, which are then real types of "practical" types (or of collective presentations).

In the interpretive understanding of the discursive interview, the contrasting reconstruction of the collective representations leads directly to the formation of real types in the sense defined above: as empirical type concepts cleansed of idiosyncrasies. The totality of the real types thereby represents and structures "only" the collective representations (real type set) that can be found in the data material (or reconstructed from it). Such a reconstruction of all real types of a reference problem usually presupposes that the sampling has already achieved the best possible (typo)logical representation of the field (cf. Sect. 3.1).

There can therefore be no separate step towards the formation of real types in analysis of collective representations as it is understood here. This is also not in the sense of a "second degree" (or on a different level than that of the text material). Thus, contrastive interpretations do require multiple and also repeated comparisons in order to arrive at the necessary groupings of derivations; however, there is no "qualitative leap" from simple comparisons to contrastive real-type formation in this process. Rather, the grouped derivations are the gradually evolving result of repeated contrastings. These are only terminated when the groupings of individual derivations have been "meaning-reconstructed", i.e. when they are more than empirical accumulations of features because they reveal a shared structure of meaning, and when further contrasting is no longer possible or promises no new insights.

It is very likely that the groupings thus obtained can be interpreted in a meaningful way, i.e. that they are comprehensibly similar and distinguishable from others: for just as the groupings of derivations gradually emerge through contrasting, so too does the understanding of their possible meanings develop with the gradually reconstructed structures in the process of contrasting. In short, the two prerequisites for the construction of real types – the grouping (of derivations) and their interpretation– are thus parallel processes.

Sets of real types and – depending on the research interest and subject matter – single real types can certainly represent satisfactory research results. Nevertheless, even for an analysis of collective representations, the development of ideal types and a typology of the collective representations of a reference problem that is as complete as possible is a desirable (if not necessary) goal. In this case, the formation of real types can be seen as an intermediate step towards the development of an (ideal) typology.

If it was pointed out earlier that ideal types can also be obtained theoretically or quantitatively, then this must be put into perspective again here for the discursive interview and the analysis of collective representations. Two considerations are crucial here:

On the one hand, it is assumed that collective representations, precisely because they are themselves typified forms of knowledge, can only be reconstructed empirically. An ideal-typical construction of collective representations types that is not based on empirical reconstructions of real-typical collective representations (or other empirically reconstructed material) is perhaps not impossible, but in any case not very promising, since it could hardly do justice to the empirical complexity of the subject area. Moreover, the position of empirically grounded theorizing (Glaser and Strauss 1967), advocated in particular by grounded theory, is shared here. "Translated" for ideal types, it is thus to be claimed that empirically based ideal types, i.e. those obtained through intensive observation of social reality, are "better" (more tenable) than those for which this is not the case.

This notwithstanding, the quality of an ideal type construction depends only partly on its empirical foundation, and in no case can ideal types be reconstructed exclusively interpretatively: The construction of interpretive ideal types is not a purely empirical procedure. They do not emerge solely through continued interpretations or even through a simple summation of characteristics. Rather, the construction of ideal types requires, in addition, a considerable amount of "theory work". The construction of one or more ideal types is therefore not a pure reconstruction process in which – however laborious – existing structures are uncovered. It is to a large extent a construction process in which the conceptual terminology gradually distances itself from the empirical object area and, in the case of an empirical-qualitative foundation, also from the real types.

The goal here is an "unrealization" or liberation from concrete object references. Only when the types are "ideal" and "empirically empty" in this sense one can meaningfully speak of a construction ideal type . For the analysis of collective representations, this means above all that the ideal types no longer contain any elements that still directly refer to the concrete object of investigation. For this process of ideal type construction, primarily mental-creative achievements are required (among other things, theoretical extrapolation and generalization, abstraction, thought-experimental variation).[30]

This perhaps also shows the difficulty mentioned above in representing the process of ideal type construction. For it is not a methodological step for which one could present corresponding procedures. One can stick to Weber's very brief and vague suggestions here, such as the thought-experimental continuation of single components (or the thematic-logical purification) of a type (Weber 1985, p. 4 f.), or orient oneself to more far-reaching interpretations or suggestions, such as those introduced into the methodological discussion by Gerhardt (1984, 1986) in particular. As stimulating and helpful as these advices for ideal type construction may be: they do not change the fact that the decisive impulse for the construction of ideal types – and even more so for that of typologies – is a mental (conceptual, theoretical) one and can therefore neither be "learned" nor "applied".

Whether, finally, a typology developed from empirical-qualitative research is good (or "better") can, in turn, neither be "proven" in general nor demonstrated for the concrete research case. For, as Weber emphasized, the value of ideal types and typologies only becomes apparent in the test of experience of later applications, namely in whether they are useful as heuristic instruments. The ideal types must also prove useful in further research and for theory-building, and in other social fields as well. However, the empirical material from which the ideal types were developed can be seen as the first and most important application case. If it is possible to achieve an better understanding of the research object with the ideal types (than with the real types), this is at least a good argument for the plausibility and viability of the developed ideal types of collective representations.

[30] This may sound more abstract and complicated than it is in actual interpretation and analysis processes. A purely fictitious example: In a study of collective representations of unemployment (cf. also the brief description by Brenke and Peter 1985 in Sect. 1.3), real types such as "unemployed people are lazy", "unemployment can affect anyone", "unemployment is normal in a globalised world", etc. could be the (subject-related) result of interpretative-reconstructive analyses. The consequence of further analytical confrontations with these real types could then be more abstract and consistent ideal types such as perhaps "victimization of social situations" or "economic fatalism". Such ideal types would then generally not be exactly matched by any single derivation (of unemployment). In order to prove useful, however, they would have to be able both to "better" explain the field under investigation and to be successfully applied to other subject areas (e.g. collective representations of other social or individual situations such as poverty, education or health).

Ensuring the Quality of Discursive Interviews

5

In Sect. 4.4, typologising was described as a way of generalising and assuring the quality of qualitative research. However, it is still unclear what exactly is meant by this. This will now be clarified in more detail in this concluding chapter. As in earlier chapters, this requires a general approach to the topic of quality assurance in qualitative social research before the associated questions can be discussed with regard to the discursive interview.

In the field of qualitative social research, different strategies can be observed with which attempts are made to underpin the quality of empirical research (as a process and as a result) and thus to increase its authority.[1] Basically, two central aspects can be distinguished:

On the one hand, the question must be answered why a recipient should "buy" the researcher's research results. This is about the *credibility of the research results*. (How does the researcher arrive at his/her interpretations? Would others arrive at the same?) In empirical social research, this question is classically assessed on the basis of quality criteria (validity, reliability, objectivity).

If research results are recognised as credible (valid), the further question then arises as to whether they are significant beyond the concrete object of investigation. What is at stake here, then, is the *generalizability* (or the degree of generalizability) of research results that are in principle recognized as credible.

[1] In this context, "empirical research" itself represents only one mode among others with which the validity of scientific work is justified (Reichertz 1999, p. 320 ff.). Among the alternative modes, Reichertz includes the charisma of the researchers, specific procedures, and scientific discourse (1999, p. 334 ff.).

C. G. Ullrich, *The Discursive Interview*,
https://doi.org/10.1007/978-3-658-38477-7_5

With regard to both questions, qualitative social research is exposed to suspicions that are difficult to refute: In the first case, there is the suspicion of subjectivism: interpretations are always and necessarily subjective; many other researchers would therefore have interpreted the data differently and come to different results. In the second case, the lack of statistical representativeness or the small number of cases, which do not permit any statistical generalisations, are criticised. As already explained in Sect. 3.1, this is based on a narrow understanding of representativeness and generalisability, which is, however, extremely widespread in the social sciences and also in "non-scientific" circles. Qualitative social research therefore often finds it difficult to convincingly demonstrate the value of its own research results to the outside world.[2]

However, qualitative social researchers largely agree that answers must be found for both questions (about credibility and generalizability). There is no agreement, however, on what these answers might look like and whether answers can be found for qualitative social research as a whole or only for single methods and approaches.

Flick (2005, p. 192 ff.) attributes the great heterogeneity in dealing with quality standards within qualitative social research primarily to the different and sometimes contradictory methodological schools as well as to different developments and traditions in language areas (especially between the Anglo-American and the German discussion) and disciplines. As a result, this has led to a discrepancy between the claim to general standards that apply to qualitative social research as a whole and "local" proposals for solutions that are limited to individual methods and areas of application. On the other hand, the question is repeatedly raised whether there can be equally valid quality criteria for all directions of qualitative social research at all, or whether a "standardization of quality standards" is even necessary or desirable (cf. e.g. Lüders 2003; Reichertz 1997; Strübing et al. 2018).

Overall, the idea of universally valid qualitative quality standards is rarely advocated (Seale 1999). Far more frequently, it is assumed that the different methodological approaches differ so fundamentally, especially with regard to their epistemological premises, that at best the formulation of relatively "binding" (consensual) quality criteria would be possible at the level of specific methodologies (cf. among others Creswell and Miller 2000; Guba and Lincoln 1998; Madill et al. 2000; Metcalfe 2005; Terhart 1995).

[2] One possible reason for this, however, could be a lack of self-criticism in dealing with questions of quality control in qualitative research.

5.1 Credibility

In order to substantiate or plausibilise the credibility of research results, so-called quality criteria are used in large parts of empirical research. Usually, a distinction is made between validity criteria (to check the validity of the results), reliability criteria (to check the reliability of a method) and objectivity criteria (independence of the research results from the researcher). The first two criteria in particular have a clearly "test and statistical" background and are not easily transferable to qualitative research.

Basically, three clearly distinguishable ways of dealing with quality criteria can be identified in qualitative research: the rejection of quality criteria, attempts to apply or adapt the classical criteria, and the development of specific criteria for qualitative methods.

The use of quality criteria is rejected primarily on the basis of two arguments: First, reference is made to the character of qualitative research. Since it serves exploration and hypothesis generation, the question of the validity of the research results is at least of secondary importance. Specific quality criteria directly aimed at the generation of research results are therefore *not necessary*. Often, there is no need for quality criteria because the validity of the research results is already sufficiently guaranteed by the analytical procedure used. On the other hand, it is claimed that the application of quality criteria is *not possible* in qualitative social research, which is mainly justified by the interpretative character of qualitative research. Interpretations are therefore per se beyond the reach of "objectified" quality criteria and external assessments.[3]

A rejection of quality criteria is objected by the postulation that all research must accept questions about its quality and must specify criteria by which third parties can judge this quality. For even if one conducts explorative or descriptive research, this can be done better or worse. The same applies to the empirically based generation of hypotheses and social science interpretation procedures, for which other (testable) requirements must apply than for artistic ones, for example.

The second basic stance assumes that the classical quality criteria of validity, reliability and objectivity are universal (i.e. not specific to a particular type of method) and can therefore also be applied to qualitative research. However, the

---

[3] In the radical constructivist version of this view, it is also argued that it is fundamentally impossible to distinguish, for example, between "good" and "bad" interpretations, which already makes the idea that one can judge research results on the basis of quality criteria seem absurd (cf. among others Lincoln and Guba 1985, p. 289 ff.; Seale 1999, p. 32 ff.; Steinke 1999, p. 131 ff.).

application or reformulation of the classical quality criteria for qualitative research (cf. among others Kirk and Miller 1986; Kelle et al. 1993; Kvale 1995; Madill et al. 2000) proves to be exceedingly difficult, albeit this claim is not completely abandoned (cf. Flick 2005, p. 195 f.).

A possible reliability criterion is, for example, the parallel but independent coding of a text passage by at least two coders. If the codes assigned largely agree, the coding process can be assessed as reliable (for examples see Madill et al. 2000; Silverman 1997). However, validity has received far more attention than the criterion of reliability[4] in the methodological discussion of qualitative social research. On the one hand, it is claimed that qualitative survey procedures, due to their openness, "naturalism" and contextualisation, among other things, already guarantee an "ecological validity" (cf. Kvale 1995; Mühlfeld et al. 1981, p. 346 ff.), i.e. a significantly higher validity than standardised procedures. Like other proposals for the validation[5] of research, however, this is not a validity criterion in the classical sense, but a general strategy for establishing validity (Mishler 1990, p. 419 f.). This also applies to "communicative validation" (member check), which can be classified as a social procedure (see below).

However, the strategy most often associated with and practiced in large parts of qualitative social research in the hope of establishing plausible validity is probably *triangulation* (cf. Denzin 1970; Flick 2008, among others). A triangulation is "the combination of methodologies in the study of the same phenomenon" (Denzin 1970, p. 291), usually referring to the combination of different types of data or methods.[6] In the case of method triangulation, the methods used may be equal or have a hierarchical relationship. If qualitative methods are combined with standardised ones, this is usually referred to as "mixed methods" (cf. Bryman 2006; Tashakkori and Teddlie 2003, among others).

---

[4] Thus Lamnek's assertion that reliability as a measurement-theoretical term "only appears in the methodological discussions of the qualitative paradigm (…) in a negatively exclusionary way" (Lamnek 1988, p. 161) is certainly exaggerated in view of some counter-examples, but it probably reflects a widespread attitude.

[5] Perhaps the best-known proposals for qualitative validation procedures are communicative, pragmatic, procedural, argumentative and cumulative validation (cf. among others Flick 2005, p. 197 f.; Kvale 1995; Lamnek 1988, p. 151 ff.). Furthermore, validity criteria and forms of validation have been proposed on the basis of very different epistemological positions, which often have hardly any recognizable relation to the classical understanding of validity (cf. among others Lather 1993; Lewis 2009; Maxwell 1992; Onwuegbuzie and Leech 2007).

[6] In addition, Denzin (1970) distinguishes researcher triangulation and theory triangulation.

Originally, triangulation was developed primarily to validate research results: If two methods arrive at the same result, so the idea goes, they validate each other or the matching result. However, such claims have not been sustained methodologically. Thus, even two or more matching results are no guarantee that the results are valid. Most importantly, it is unclear how to deal with the situation when the results of different methods lead to contradictory results. Today, triangulative methods are therefore also seen more as alternative forms of validity justification (and not as a means of validation) (cf. Flick 2008, p. 17 ff.).

The difficulties in "translating" classical quality criteria for qualitative contexts have tended to increase doubts about their applicability and appropriateness. However, if one does not want to dispense with quality criteria altogether, the only way out is to *develop and apply special criteria that are appropriate to qualitative social research as* a whole or to single approaches and methods. The proposal of Lincoln and Guba, who suggest credibility, transferability, reliability, confirmability, and authenticity as specific qualitative quality criteria (or criteria for establishing trustworthiness), can be considered classic here (Lincoln and Guba 1985, p. 289 ff.; Guba and Lincoln 1998, p. 213).

Social Procedures of Justification of Validity

Of particular importance for the discussion about quality standards in qualitative social research are procedures that aim to establish validity via social processes. In this context, the quality of the research is to be secured through its acceptance by relevant reference groups.[7] In principle, a distinction must be made here between validity evaluation by the "researched" and in scientific discourse.

The first variant is known in the German-speaking context primarily as *communicative validation* (also "member check" or "member validation"). Here, the research process and/or the research results are "checked" by the researched (cf. Scheele and Groeben 1988). The influence of the researched can range from a mere commentary on research results to active participa-

(*continued*)

[7] Ultimately, the validity of empirical research results always depends on social processes, namely on whether research results are recognized or adequately appreciated by the "scientific community". In contrast to the criteria discussed above, however, the discursive form of the justification of validity is consistently understood as the decisive, if not the only, yardstick in the processes described as "social".

(continued)

tion in, if not control of, the research process (Bloor 1997, p. 41 ff.). Communicative validation is controversially discussed in qualitative social research. Especially for broad areas of ethnography and particularly in so-called action research, it is a central research tool whose importance goes far beyond controlling validity. In contrast, most qualitative researchers reject it as an illegitimate restriction of the researcher (Flick 2005, p. 197).

The second form understands the justification of validity as a social discourse, which is conducted in particular in the "scientific community".[8,9] Above all, this includes procedures known as *"peer examination"*, in which an external examination of single research steps is carried out. The type and scope of a "peer examination" can also vary greatly, ranging from simple exchanges to extensive reviews in auditing processes (cf. Lincoln and Guba 1985, pp. 318 ff., 382 ff.). There is a large number of individual proposals as well as comprehensive catalogues of criteria for social procedures for reaching validity and especially for "peer examination". A central aspect here is the creation of transparency and intersubjective verifiability through detailed documentation of the entire research process (Steinke 1999, p. 208 ff.).

In the meantime, many such proposals for specific qualitative quality criteria have been made (cf. among others Huberman and Miles 1998; Mishler 1990; Steinke 1999). Most of them propose "soft" criteria such as the "appropriateness" of the method (Steinke 1999, p. 215 ff.), the documented reflexivity of the researchers (Seale 1999, p. 159 ff.; Steinke 1999, p. 231 ff.; see also Sect. 2.2.3) and, above all, social procedures for establishing validity (see Box). Critical objections to such proposed criteria for assessing qualitative social research are that the individual proposals are usually very general, too vaguely formulated and therefore difficult to apply. Furthermore, many proposals seem appropriate only for individual methods and therefore cannot claim to be generally valid quality criteria for

---

[8] In this sense, Mishler (1990, p. 417) defines validation as the "social construction of knowledge", in which the decisive factor is whether "the relevant community of scientists evaluates reported findings as sufficiently trustworthy to rely on them for their own work". In a broader sense, this also includes procedural steps that provide for mutual control of researchers already in the research process itself (e.g. interpretation in groups).

[9] Other reference groups that can be considered include the public, clients, addressees, or "practice" in general. The "practical relevance" of research is often seen as a quality criterion in its own right (cf. e.g. Hammersley 1992; Steinke 1999, p. 241 ff.), if it is not even elevated to the central quality characteristic of research in general – as in action and participatory research.

qualitative social research. Despite the many proposals, it has therefore not yet been possible to develop binding (or even halfway consensual) quality criteria for qualitative social research.

Since no convincing solutions have yet been found, none of the three basic positions on validity outlined here has yet been able to establish itself. It also seems clear that many qualitative social researchers think only little about the justification of validity, or only in relation to their specific research. At the same time, there are apparently something like reference points and implicit rules with which the researchers signal the quality their research. These include the use of established and increasingly "standardized" and codified methods, a high degree of "regulation" of these methods (or the tendency to use methods with strong rules), and the disclosure of the methodological procedure (among other things: precise presentation of the individual work steps; examples of interpretation; offensive discussion of research problems).

5.2 Scope

In addition to the credibility of research results, their generalizability (generalization) is considered the second important quality criterion for the quality of research. The question of whether and to what extent research findings can claim relevance beyond the immediate object of research presupposes (positive) credibility. (If research findings are not credible, then there is no reason to generalize them.) However, research results do not necessarily have to be generalizable (especially in applied research). But with the degree of their scope or generalizability, the significance and value of research results also increase (ceteris paribus).

For this second form of judging research quality, too, the answers within qualitative social research are very contradictory. First of all, there is also the view that a generalisation of qualitative research findings is *not possible* or *not necessary*. On the one hand, the impossibility of (statistical) generalizations can be regretted as an unavoidable price to be paid for a description that is close to the subject and "thick", for example, when the interest is in complex individual phenomena (Mayntz 1985). On the other hand, generalization can be considered unnecessary with regard to the goal of generating hypotheses or empirically grounded theory building. After all, how common discovered or reconstructed correlations are is a question that actually stands outside of an exploratory or hypothesis-generating research interest, or at least is secondary here. On the whole, questions of generalizability (similar to the classical quality criteria) are closely linked to the logic of quantitative procedures and statistical measurement and are therefore rejected as inappropriate for qualitative research (cf. among others Geertz 1987, p. 30 ff.; Lincoln and Guba 1985, p. 110 ff.).

However, the claim to generalisation is also accepted in principle in large parts of qualitative social research and is sometimes even aggressively advocated (e.g. Williams 2000). At the very least, various strategies exist to increase the scope of empirical findings or to make them more plausible. A rather pragmatic strategy of generalization consists in achieving a stronger generalizability of qualitative findings by linking them with standardized methods (*mixed methods*) (cf. among others Flick 2008, p. 108 f.; Hammersley 1992, p. 86 ff.; Seale 1999, p. 113 ff.), i.e. indirectly via the "detour" of statistical representativeness. A quite different strategy is *theoretical generalization*. These are attempts to extend the scope of research findings by linking them to theories (see, among others, Becker 1998, p. 146 ff.; Hammersley 1992, p. 91 ff.; Przyborski and Wohlrab-Sahr 2014, p. 366 f.; Seale 1999, p. 109 ff.).

Despite these and other attempts to ensure generalizability even in the context of qualitative research, *generalization-skeptical positions* dominate the qualitative methodological literature ("The only generalization is: there is no generalization"; Lincoln and Guba 1985, p. 110). Insofar as this does not lead to a simple rejection of the claim to generalization, strategies are proposed with which a validity that is not limited to the immediate context of investigation could be justified in a form appropriate to qualitative social research.

Proposals for original qualitative means of increasing the range (or applicability) of qualitative research results include the "transferability" of results to other contexts (Lincoln and Guba 1985, p. 110 ff.) and "limitation" (Steinke 1999, p. 27 ff.). Similar to the criteria for credibility, these proposals are also very vague and unclear with regard to their application (Seale 1999, p. 107 f.). Moreover, these proposals differ from equally "soft" concepts such as theoretical generalization less in terms of the procedure than in that the objective is not referred to as generalization.

The situation in qualitative social research is thus heterogeneous and controversial also with regard to the question of generalisation. On the other hand, there have long been two qualitative "grand strategies" for increasing the scope of qualitative research results, which have already been presented elsewhere here (cf. Sects. 3.1 and 4.4).

On the one hand, this is the formation of *typologies* (cf. Kluge 1999; Weber 1995 [1922]), which are not generalizations in the sense of statistical representativeness, but can have a generalization-equivalent function. For at least one task of typologies is also to depict the field of the phenomenon under study in structural-typological (rather than quantitative-proportional) terms. In this way, conclusions can then be drawn for the entire subject area.

The second important strategy in this context is "*theoretical sampling*", as described by Glaser and Strauss as a fundamental procedure of "grounded theory" (Glaser and Strauss 1967). Although the aim of theoretical sampling is to enable the development of object-related theories, it can nevertheless also be regarded as a fundamental alternative to procedures of "representativistic" generalisation. This is because "theoretical sampling" aims at representation and theoretical saturation already in the research process itself through *systematic variation*, which makes efforts at ex post generalization appear superfluous (cf. Glaser and Strauss 1967; Gobo 2004; Seale 1999, p. 87 ff.).

5.3 Quality Control in Collective Representations Analysis with Discursive Interviews

In principle, an unagitated approach to questions of quality is to be recommended, although there should be no doubt about its fundamental necessity. In view of the almost insurmountable "transferability problems", however, it is equally beyond doubt that neither the classical quality criteria nor a generalisation claim in the sense of statistical representativeness are suitable for assessing the quality of reconstructions of collective representations. However, this in turn does not change the questions or doubts about credibility formulated at the outset ("Are these not just subjective interpretations?") and about the scope of research results ("Does this not only apply to the object studied?"). Both questions directly address the quality of the research and should therefore also be answerable for analysis of collective representations based on discursive interviews.

The Discursive Interview sees itself as a comprehensive qualitative research method in which the individual procedural steps are integrated and oriented towards the goal of reconstructing collective representations. Quality control in the Discursive Interview therefore always takes place in and through the research process itself. An ex post examination in the form of one or more separate procedures following the research process is therefore not necessary – neither for the question of credibility nor for that of the scope of the research results.

Quality aspects have already been explained accordingly during the presentation of the methodological procedure. Central here is the comprehensive and (hopefully) convincing methodological justification of all research steps. In the case of the discursive interview, these are above all the sampling process, the form of the interview, the methodological foundation of question forms, the contrasting interpretation and the type formation. At this point, these aspects of research quality need only be "remembered" and brought together. (Nevertheless, "complementary measures" are suggested at the end of this section for an additional ensuring of quality).

According to the view taken here, generalization is often not necessary, not possible, and perhaps not even desirable for questions investigated using qualitative methods. At the same time, it can be formulated as a basic rule that any research has little to object to if the *scope of* the research results extends beyond the narrower research context. In short, if research results are not or hardly generalizable, they can still be good. However, their value increases with their scope.

This also applies to the analysis of collective representations with discursive interviews. The reconstruction of individual representations as well as sets of real types close to the subject matter are, if they succeed convincingly, meaningful and satisfying research results. An ideal typology – in the case of collective representations, that is, a general typology of collective representations – is nevertheless to be regarded as a higher-quality result because it is independent of concrete contexts. Therefore, such a typology should always be the goal, at least the "long-term objective", of a collective representations analysis. As explained in Sect. 4.4, however, this cannot be firmly planned for because the empirical material is often insufficient for the formation of ideal types.

Ideal types and typologies can be regarded as a specific form of generalisation. One should be aware, however, that this form of generalization has little in common with the widespread understanding of generalization as statistical representativeness. It therefore seems less misleading to describe the formation of real and ideal types as strategies for increasing the scope or, even more generally, for increasing the value of qualitative research results.

An analysis of collective representations with discursive interviews does *not* initially have *any credibility or authenticity problem*, i.e. at the level of discovering individual derivations. For individual derivations captured with interviews are, as explained in Sect. 2.1, nothing other than collective representations applied in communications. As such, they are always "authentic" and cannot be "false" or "not valid".[10]

However, this only applies to the respective derivations themselves, but not to the reconstruction of derivations and collective representations in the research process. For derivations and collective representations can possibly only be reconstructed incompletely. This applies both to individual representations and to the totality of the collective representations of a reference problem.

The reasons for this can vary, but are mainly due to three "sources of error".

[10] On the other hand, it is possible that interviewees deceive themselves or the interviewers about the actually effective individual representations and the "true" motives of their actions. However, these are not the subject of an analysis of collective representations by means of discursive interviews.

- Thus, insufficient interviewing can lead to the fact that justifications and other texts containing derivations are not available to a sufficient extent. The number of individual derivations would then not be sufficient for a reconstruction of collective representations, or the reconstruction of collective representations would take place on a "too thin" basis. This would lead to a problem of credibility with regard to the reconstructed collective representations.
- A similar effect can be caused by an insufficient (e.g. too early terminated) "theoretical sampling". The then missing contrasting options and a possibly too small number of cases already complicate the reconstruction of derivations and collective representations, but will have a detrimental effect above all on the formation of types.
- Irrespective of "theoretical sampling", too little or "wrong" contrasting and grouping (e.g. as a result of inadequate coding) can also significantly impair, if not prevent, type formation and thus significantly reduce the (possible) scope of the research results.

Since such "mistakes" – often they will rather be consequences of conflicting goals or an "awkward" empirical object – cannot be prevented in principle, two general rules should also be observed for an analysis of collective representations by means of discursive interviews, which can additionally ensure the quality of the research.

The first consists in the constant and repeated linking back or checking of all evaluation and interpretation steps against the data material. This is a basic principle of interpretative reconstruction of collective representations (cf. Sect. 4.2). Beyond the necessary use within the framework of interpretative reconstructions, however, such checks on the data material can also be used for the (self-)control of the interpretation process (e.g. if this is done at certain, previously determined points in time). In addition, such tests can also be used specifically for the "verification" (hardening) of reconstructions that have already been carried out, in particular by applying them to data material that has not yet been evaluated.

The second general rule for additional quality assurance is to document the entire research process as comprehensively as possible and in a way that is accessible to the public. In addition to the methods (discursive interview, contrasting reconstruction), this includes above all practical research decisions (especially in the case of problems in the field) as well as the disclosure or provision of the data material (cf. also Steinke 1999, p. 208 ff.). Detailed documentation of the entire procedure is, on the one hand, of course a prerequisite for ensuring that the backward linkage of the interpretation steps can also be understood "from the outside". But it also has an intrinsic value that goes beyond this comprehension. This is because such documentation (including the problems and solutions encountered during the research) generally increases researcher honesty. In extreme cases, every single step could be "replicated" and checked by uninvolved parties, and at no point would it be left in the dark "how they came up with it".[11]

These two rules should be followed primarily for "self-check", especially in order to identify any weaknesses in one's own research at all. At the same time, however, they also enable at least partial external inspection and thus contribute to the credibility and scope of the research results. Finally, these can also be enhanced by a triangulative approach. In particular, group discussions and document analyses (of both classic material and Internet sources) can be used to reconstruct collective representations and can be used in addition to discursive interviews. Insofar as the results of the different methodological approaches and types of data support, supplement, modify or differentiate the reconstructions from the discursive interviews, credibility can be increased in this way – albeit beyond narrower ideas of validation and generalisation – and the claim to validity beyond the immediate research object can be substantiated.

[11] However, such "ideal" documentation is unlikely: on the one hand, because of the extremely high workload; on the other hand, however, primarily because of data protection and research ethics limitations.

References

Adamzik, Kirsten. 1995. *Textsorten – Texttypologie. Eine kommentierte Bibliographie*. Münster: Nodus.

———. 2008. Textsorten und ihre Beschreibung. In *Textlinguistik. 15 Einführungen*, ed. Nina Janich, 145–175. Tübingen: Narr.

Alemann, Annette von. 2014. Geschlechterungleichheit in Führungspositionen der deutschen Wirtschaft: Ergebnisse einer Deutungsmusteranalyse. *Gender – Zeitschrift für Geschlecht, Kultur und Gesellschaft* 6 (3): 10–25.

———. 2015. *Gesellschaftliche Verantwortung und ökonomische Handlungslogik. Deutungsmuster von Führungskräften der deutschen Wirtschaft*. Wiesbaden: Springer VS.

Alvesson, Mats. 2011. *Interpreting interviews*. Los Angeles: Sage.

Antaki, Charles. 1994. *Explaining and arguing. The social organization of accounts*. London: Sage.

Arnold, Rolf. 1983. Deutungsmuster. Zu den Bedeutungselementen sowie den theoretischen und methodologischen Bezügen eines Begriffs. *Zeitschrift für Pädagogik* 29: 893–912.

Auer, Peter. 1996. From context to contextualization. *Links & Letters* 3: 11–28.

Baker, Carolyn D. 2002. Ethnomethodological analyses of interviews. In *Handbook of interview research. Context and method*, ed. Jaber F. Gubrium and James A. Holstein, 777–795. Thousand Oaks: Sage.

Barton, Allen H., and Paul F. Lazarsfeld. 1979. Einige Funktionen von qualitativer Analyse in der Sozialforschung. In *Qualitative Sozialforschung*, ed. Christel Hopf and Elmar Weingarten, 41–89. Stuttgart: Klett-Cotta (Erstveröffentlichung 1955).

Bastian, Jasmin. 2014. *Väter und das Vorlesen. Eine Deutungsmusteranalyse*. Wiesbaden: Springer VS.

Becker, Howard S. 1959. *Soziologie als Wissenschaft vom sozialen Handeln*. Würzburg: Holzner.

———. 1998. *Tricks of the trade. How to think about your research while you're doing it*. Chicago: University of Chicago Press.

C. G. Ullrich, *The Discursive Interview*,
https://doi.org/10.1007/978-3-658-38477-7

Behnke, Cornelia; Meuser, Michael 1997: Zwischen aufgeklärter Doppelmoral und partnerschaftlicher Orientierung. *Frauenbilder junger Männer. Zeitschrift für Sexualforschung* 10: 1–18.

Bellah, Robert N., Richard Madsen, Willam M. Sullivan, Ann Swidler, and Steven M. Tipton. 1985. *Habits of the heart*. In *Individualism and commitment in American life*. Berkeley: University of California Press.

Belnap, Nuel D., and Thomas B. Steel. 1976. *The logic of questions and answers*. New Haven: Yale University Press.

Berger, Peter L., and Thomas Luckmann. 1990. *Die gesellschaftliche Konstruktion der Wirklichkeit. Eine Theorie der Wissenssoziologie*. Frankfurt a. M.: Fischer (Erstveröffentlichung 1966).

Bergmann, Jörg R. 1981. Frage und Frageparaphrase: Aspekte der redezuginternen und sequenziellen Organisation eines Äußerungsformats. In *Methoden der Analyse von Face-to-Face-Situationen*, ed. Peter Winkler, 128–142. Stuttgart: Metzler.

Blommaert, Jan. 2001. Context is/as critique. *Critique of Anthropology* 21 (1): 13–32.

Bloor, Michael. 1997. Techniques of validation in qualitative research a critical commentary. In *Context and method in qualitative research*, ed. Gale Miller and Robert Dingwall, 37–50. London: Sage.

Blumer, Herbert. 1973. Der methodologische Standort des symbolischen Interaktionismus. In *Alltagswissen, Interaktion und gesellschaftliche Wirklichkeit. Bd. 1 "Symbolischer Interaktionismus und Ethnomethodologie"*, ed. Arbeitsgruppe Bielefelder Soziologen, 80–146. Reinbek: Rowohlt.

Bögelein, Nicole. 2016. *Deutungsmuster von Strafe. Eine strafsoziologische Untersuchung am Beispiel der Geldstrafe*. Wiesbaden: Springer VS.

Bögelein, Nicole, and Nicole Vetter. 2019. *Der Deutungsmusteransatz: Einführung – Erkenntnisse – Perspektiven*. Weinheim: Beltz Juventa.

Bogner, Alexander, and Wolfgang Menz. 2001. "Deutungswissen" und Interaktion. Zu Methodologie und Methodik des theoriegenierenden Experteninterviews. *Soziale Welt* 52 (4): 477–500.

Bogner, Alexander, Beate Littig, and Wolfgang Menz, eds. 2002. *Das Experteninterview. Theorie, Methode, Anwendung*. Opladen: Leske + Budrich.

Bohnsack, Ralf. 1989. *Generation, Milieu, Geschlecht. Ergebnisse aus Gruppendiskussionen mit Jugendlichen*. Opladen: Leske + Budrich.

———. 1991. *Rekonstruktive Sozialforschung. Einführung in Methodologie und Praxis qualitativer Forschung*. Opladen: Leske + Budrich.

———. 1992. Dokumentarische Interpretation von Orientierungsmustern. Verstehen – Interpretieren – Typenbildung in wissenssoziologischer Analyse. In *Analyse sozialer Deutungsmuster: Beiträge zur empirischen Wissenssoziologie*, ed. Michael Meuser and Rainer Sackmann, 139–160. Pfaffenweiler: Centaurus.

———. 1997. "Orientierungsmuster". Ein Grundbegriff qualitativer Sozialforschung. In *Methodische Probleme der empirischen Erziehungswissenschaft*, ed. Folker Schmidt, 49–61. Baltmansweiler: Schneider.

———. 2000. Gruppendiskussion. In *Qualitative Forschung. Ein Handbuch. Reinbek*, ed. Uwe Flick, Ernst von Kardorff, and Ines Steinke, 369–384. Reinbek: Rowohlt.

Bohnsack, Ralf, Nentwig-Gesemann Iris, and Nohl Arnd-Michael, eds. 2007. *Die dokumentarische Methode und ihre Forschungspraxis. Grundlagen qualitativer Sozialforschung*. 2nd ed. Wiesbaden: VS Verlag.

Bosancic, Sasa. 2014. *Arbeiter ohne Eigenschaften. Über die Subjektivierungsweisen angelernter Arbeiter*. Wiesbaden: Springer VS.

Bourdieu, Pierre. 1976. *Entwurf einer Theorie der Praxis auf der ethnologischen Grundlage der kabylischen Gesellschaft*. Frankfurt a. M: Suhrkamp.

Bourdieu, Pierre, et al. 2002. *Das Elend der Welt. Zeugnisse und Diagnosen alltäglichen Leidens an der Gesellschaft*. Konstanz: UVK.

Brenke, Karl, and Michael Peter. 1985. Arbeitslosigkeit im Meinungsbild der Bevölkerung. In *Gewandelte Werte, erstarrte Strukturen. Wie Bürger Wirtschaft und Arbeit erleben*, ed. Michael von Klipstein and Burkhard Strümpel, 87–127. Bonn: Verlag Neue Gesellschaft.

Breuer, Franz. 2010. Wissenschaftstheoretische Grundlagen qualitativer Methodik in der Psychologie. In *Handbuch Qualitative Forschung in der Psychologie*, ed. Günther Mey and Katja Mruck, 35–49. Wiesbaden: VS Verlag.

Breuer, Franz, Günter Mey, and Katja Mruck. 2011. Subjektivität und Selbst-/Reflexivität in der Grounded-Theory-Methodologie. In *Grounded theory reader*, ed. Günter Mey and Katja Mruck, 2nd ed., 427–448. Wiesbaden: VS Verlag.

Breuer, Franz, Petra Muckel, and Barbara Dieris. 2018. *Reflexive Grounded Theory. Eine Einführung für die Forschungspraxis*. Wiesbaden: Springer VS.

Briggs, Charles. 1986. *Learning how to ask. A socio-linguistic appraisal of the role of the interviewer in social science research*. Cambridge: Cambridge University Press.

Brinker, Klaus. 2005. *Linguistische Textanalyse. Eine Einführung in die Grundbegriffe und Methoden.*, 6. überarbeitete u. erweiterte Aufl. Berlin: Schmidt.

Brinkmann, Svend. 2007. Could interviews be epistemic? An alternative to qualitative opinion polling. *Qualitative Inquiry* 13 (8): 1116–1138.

Bryman, Alan. 2006. Integrating quantitative and qualitative research. How is it done? *Qualitative Research* 6 (1): 97–113.

Bühler, Karl. 1999. *Sprachtheorie: Die Darstellungsfunktion der Sprache*. 3rd ed. Stuttgart: Fischer.

Charmaz, Cathy. 2014. *Constructing grounded theory*. 2nd ed. Los Angeles: Sage.

Cicourel, Aaron V. 1974. *Methode und Messung in der Soziologie*. Frankfurt a. M.: Suhrkamp (Erstveröffentlichung 1964).

———. 1982. Interviews, survey, and the problem of ecological validity. *The American Sociologist* 17: 11–20.

Cicourel, Aaron. 2005. Elicitation as a problem of discourse. In *Sociolinguistics. An international handbook of the science of language and society*, ed. Ulrich Ammon, Norbert Dittmar, Klaus J. Mattheier, and Peter Trudgill, 2nd ed., completely revised and extended Aufl., 1013–1022. Berlin: de Gruyter.

Clark, Herbert H., and T.V. Carlson. 1981. Context for comprehension. In *Attention and performance IX*, ed. J. Long and A. Baddeley, 313–330. Hillside: Lawrence Erlbaum Associates.

Conrad, Rudi. 1978. *Studien zur Syntax und Semantik von Frage und Antwort*. Berlin: Akademie.

Creswell, John W., and Dana L. Miller. 2000. Determining validity in qualitative inquiry. *Theory into Practice* 39 (3): 124–130.

Danner, Helmut. 2006. *Methoden geisteswissenschaftlicher Pädagogik*. 5th ed. München: Ernst Reinhardt.

de Swaan, Abram. 1993. *Der sorgende Staat: Wohlfahrt. Gesundheit und Bildung in Europa und den USA der Neuzeit*. New York: Campus.

Denzin, Norman K. 1970. *The research act*. Chicago: Aldine.

Deppermann, Arnulf. 2013. Interview als Text vs. Interview als Interaktion [61 Absätze]. *Forum: Qualitative Social Research* 14 (3): Art. 13.

Dey, Ian. 1999. *Grounding grounded theory*. San Diego: Academic Press.

Diekmann, Andreas. 1999. *Empirische Sozialforschung. . Grundlagen, Methoden, Anwendungen*. Reinbek: Rowohlt.

Dittmar, Norbert. 2009. *Transkription. Ein Leitfaden mit Aufgaben für Studenten, Forscher und Laien*. Wiesbaden: VS.

Douglas, Jack D. 1985. *Creative interviewing*. Beverly Hills: Sage.

Duranti, Alessandro, and Charles Goodwin, eds. 1992. *Rethinking context. Language as an interactive phenomenon*. Cambridge: Cambridge University Press.

Durkheim, Emile. 1984. *Regeln der soziologischen Methode*. Frankfurt a. M.: Suhrkamp.

———. 1981. *Die elementaren Formen des religiösen Lebens*. Frankfurt a. M.: Suhrkamp (Erstveröffentlichung 1912).

Eisermann, Gottfried. 1962. Vilfredo Pareto als Wissenssoziologe. *Kyklos* 15: 427–464.

Esping-Andersen, Gøsta. 1990. *The three worlds of welfare capitalism*. Princeton: Princeton University Press.

Eucken, Walter. 1947. *Die Grundlagen der Nationalökonomie*. Godesberg: Verlag Helmut Küpper.

Fink, Arlene. 1995. *How to ask survey questions*. Thousand Oaks: Sage.

Fischer-Rosenthal, Wolfram, and Gabriele Rosenthal. 1997. Narrationsanalyse biographischer Selbstpräsentation. In *Sozialwissenschaftliche Hermeneutik*, ed. Ronald Hitzler and Anne Honer, 133–164. Opladen: Leske + Budrich.

Fix, Ulla. 2008. Text und Textlinguistik. In *Textlinguistik. 15 Einführungen*, ed. Nina Janich, 15–34. Tübingen: Narr.

Flick, Uwe. 1995. *Qualitative Sozialforschung. Theorie, Methoden, Anwendung in Psychologie und Sozialwissenschaften*. Reinbek: Rowohlt.

———. 2005. Standards, Kriterien, Strategien. Zur Diskussion über Qualität qualitativer Sozialforschung. *Zeitschrift für Qualitative Bildungs-, Beratungs- und Sozialforschung* 6 (2): 191–210.

———. 2008. *Triangulation. Eine Einführung*. 2nd ed. Wiesbaden: VS Verlag.

Foddy, William. 1993. *Constructing questions for interviews and questionnaires. Theory and practices in social research*. Cambridge: Cambridge University Press.

Fontana, Andrea. 2002. Postmodern trends in interviewing. In *Handbook of interview research. Context and method*, ed. Jaber F. Gubrium and James A. Holstein, 161–175. Thousand Oaks: Sage.

Franck, Dorothea. 1996. Kontext und Kotext. In *Sprachphilosophie. Ein internationales Handbuch zeitgenössischer Forschung*, ed. Marcelo Dascal et al., 1323–1336. Berlin: De Gruyter.

Friedrichs, Jürgen. 1973. *Methoden empirischer Sozialforschung*. Opladen: Westdeutscher Verlag.

Früh, Doris. 2000. Online-Forschung im Zeichen des Qualitativen Paradigmas. Methodologische Reflexion und empirische Erfahrungen. *Forum Qualitative Sozialforschung* 1 (3): Art. 35. http://nbn-resolving.de/urn:nbn:de:0114-fqs0003353. Accessed: 9. Mai 2014.

Gans, Herbert J. 1992. Über die positiven Funktionen der unwürdigen Armen. Zur Bedeutung der 'underclass' in den USA. In *Armut im modernen Wohlfahrtsstaat*, ed. Stephan Leibfried and Wolfgang Voges, 48–61. Wiesbaden: Westdeutscher Verlag.

Garfinkel, Harold. 1967. *Studies in ethnomethodology*. Englewood Cliffs: Prentice-Hall.

Garz, Detlef, ed. 1994. *Die Welt als Text. Theorie, Kritik und Praxis der objektiven Hermeneutik*. Frankfurt a. M.: Suhrkamp.

Geertz, Clifford. 1987. Dichte Beschreibung. Bemerkungen zu einer deutenden Theorie von Kultur. In *Dichte Beschreibung. Beiträge zum Verstehen kultureller Systeme*, ed. Clifford Geertz, 7–43. Frankfurt a. M.: Suhrkamp.

Geremek, Bronislaw. 1991. *Geschichte der Armut. Elend und Barmherzigkeit in Europa*. München: dtv.

Gerhardt, Uta. 1984. Typenrekonstruktion bei Patientenkarrieren. In *Biographie und soziale Wirklichkeit: Neue Beiträge und Forschungsperspektiven*, ed. Martin Kohli and Günther Robert, 53–77. Stuttgart: Metzler.

———. 1986. *Patientenkarrieren*. Frankfurt a. M.: Suhrkamp.

Giddens, Anthony. 1997. *Die Konstitution der Gesellschaft. Grundzüge einer Theorie der Strukturierung*. New York: Campus.

Gläser, Jochen, and Grit Laudel. 2009. *Experteninterviews und qualitative Inhaltsanalyse*. 3rd ed. Wiesbaden: VS Verlag.

Glaser, Barney G., and Anselm L. Strauss. 1967. *The discovery of grounded theory. Strategies for qualitative research*. Chicago: Aldine.

Gobo, Giampietro. 2004. Sampling, representativeness and generalizability. In *Qualitative research practice*, ed. Clive Seale et al., 405–426. London: Sage.

Goffman, Erving. 1977. *Rahmen-Analyse. Ein Versuch über die Organisation von Alltagserfahrungen*. Frankfurt a. M.: Suhrkamp.

———. 1981a. *Forms of talk*. Oxford: Basil Blackwell.

———. 1981b. Replies and responses. In *Forms of talk*, ed. Erving Goffman, 5–77. Oxford: Basil Blackwell.

———. 1982. *Das Individuum im öffentlichen Austausch. Mikrostudien zur öffentlichen Ordnung*. Frankfurt a. M.: Suhrkamp.

———. 1986. *Interaktionsrituale. Über Verhalten in direkter Kommunikation*. Frankfurt a. M.: Suhrkamp.

Goodwin, Charles, and Alessandro Duranti. 1992. Rethinking context. An introduction. In *Rethinking context. Language as an interactive phenomenon*, ed. Alessandro Duranti and Charles Goodwin, 1–42. Cambridge: Cambridge University Press.

Goody, Esther N. 1978. Towards a theory of questions. In *Questions and politeness*, ed. Esther N. Goody, 17–43. Cambridge: Cambridge University Press.

Graumann, Carl F., Alexandre Métreaux, and Gert Schneider. 1991. Ansätze des Sinnverstehens. In *Handbuch qualitativer Sozialforschung*, ed. Uwe Flick et al., 67–77. München: Psychologie Verlagsunion.

Groenemeyer, Axel. 1999. Armut. In *Handbuch soziale Probleme*, ed. Günter Albrecht et al., 270–318. Opladen: Westdeutscher Verlag.

Guba, Egon G., and Yvonna S. Lincoln. 1998. Competing paradigms in qualitative research. In *The landscape of qualitative research*, ed. Norman K. Denzin and Yvonna S. Lincoln, 195–220. Thousand Oaks: Sage.

Gubrium, Jaber F., and James A. Holstein, eds. 2012. *The Sage handbook of interview research. The complexity of the craft*. Thousand Oaks: Sage.

Gumperz, John J. 1992. Contextualization and understanding. In *Rethinking context. Language as an interactive phenomenon*, ed. Alessandro Duranti and Charles Goodwin, 229–250. Cambridge: Cambridge University Press.

Haller, Michael. 2001. *Das Interview*. 5th ed. Konstanz: UVK.

Hammersley, Martyn. 1992. *What's wrong with ethnography?* London: Routledge.

———. 2013. *What is qualitative research?* London: Bloomsbury.

Hammersley, Martyn, and Roger Gomm. 2008. Assessing the radical critique of interviews. In *Questioning qualitative inquiry. Critical essays*, ed. Martyn Hammersley. London: Sage.

Harris, Wendell. 1988. *Interpretive acts: In search of meaning*. Oxford: Clarendon.

Heinemann, Wolfgang. 2000. Textsorte – Textmuster – Texttyp. In *Text- und Gesprächsanalyse. Ein internationales Handbuch zeitgenössischer Forschung. 1. Halbband*, ed. Klaus Brinker et al., 507–523. Berlin: De Gruyter.

Helfferich, Cornelia. 2009. *Die Qualität qualitativer Daten. Manual für die Durchführung qualitativer Interviews*. Wiesbaden: VS Verlag.

Hempel, Carl. 1976. Typologische Methoden in den Sozialwissenschaften. In *Logik der Sozialwissenschaften*, ed. Ernst Topitsch, 85–103. Köln: Kiepenheuer & Witsch.

Heritage, John. 1984. *Garfinkel and ethnomethodology*. Cambridge: Polity Press.

———. 2002. Ad Hoc inquiries: Two preferences in the design of routine questions in a open context. In *Standardization and tacit knowledge. Interaction and practice in the survey interview*, ed. Douglas W. Maynard et al., 313–334. New York: Wiley.

Hermanns, Harry. 2000. Interviewen als Tätigkeit. In *Qualitative Forschung. Ein Handbuch*, ed. Uwe Flick, Ernst von Kardorff, and Ines Steinke, 360–368. Reinbek: Rowohlt.

Hester, Stephen, and David Francis. 1994. Doing data: The local organization of a sociological interview. *British Journal of Sociology* 45 (4): 675–695.

Hindelang, Götz. 1980. Zur Klassifikation der Fragenhandlungen. In *Sprache: Verstehen und Handeln. Akten des 15. Linguistischen Kolloquiums Münster 1980*, ed. Hindelang Götz and Werner Zillig, 215–225. Tübingen: Niemeyer.

———. 1994. *Einführung in die Sprechakttheorie: Sprechakte, Äußerungsformen, Sprechaktsequenzen*. Tübingen: Niemeyer.

———. 1995. Frageklassifikation und Dialoganalyse. In *Der Gebrauch der Sprache: Festschrift für Franz Hundsnurscher zum 60. Geburtstag*, ed. Götz Hindelang, Eckard Rolf, and Werner Zillig, 176–196. Münster: Lit.

Hitzler, Ronald. 1993. Verstehen: Alltagspraxis und wissenschaftliches Programm. In *"Wirklichkeit" im Deutungsprozeß. Verstehen und Methoden in den Kultur- und Sozialwissenschaften*, ed. Thomas Jung and Stefan Müller-Dohm, 223–240. Frankfurt a. M.: Suhrkamp.

Hitzler, Ronald, and Anne Honer, eds. 1997. *Sozialwissenschaftliche Hermeneutik*. Opladen: Leske + Budrich.

Hochschild, Jennifer L. 1981. *What's fair? American beliefs about distributive justice*. Cambridge: Harvard University Press.

Hoffmann-Riem, Christa. 1980. Die Sozialforschung einer interpretativen Soziologie. *Kölner Zeitschrift für Soziologie und Sozialpsychologie* 32: 339–372.

Holstein, James A., and Jaber F. Gubrium. 1995. *The active interview*. Sage University Paper Series on Qualitative Research Methods, Bd. 37. Thousand Oaks: Sage.

———, eds. 2002. *Handbook of interview research. Context and method.* Thousand Oaks: Sage.

———. 2011. Animating interview narratives. In *Qualitative research*, ed. David Silverman, 3rd ed., 149–167. London: Sage.

Honegger, Claudia. 1978. Die Hexen der Neuzeit. Analysen zur anderen Seite der okzidentalen Rationalisierung. In *Die Hexen der Neuzeit*, ed. Claudia Honegger, 21–151. Frankfurt a. M.: Suhrkamp.

———. 2001. Deutungsmusteranalyse reconsidered. In *Materialität des Geistes. Zur Sache Kultur – Im Diskurs mit Ulrich Oevermann*, ed. Roland Burkholz, Christel Gärtner, and Ferdinand Zehentreiter, 107–136. Velbrück: Weilerwist.

Hopf, Christel. 1978. Die Pseudo-Exploration. Überlegungen zur Technik qualitativer Interviews in der Sozialforschung. *Zeitschrift für Soziologie* 7: 97–115.

———. 1988. *Qualitative Interviews in der Sozialforschung. Ein Überblick.* Berlin: Max-Planck-Institut für Bildungsforschung.

———. 2000. Forschungsethik und qualitative Forschung. In *Qualitative Forschung. Ein Handbuch*, ed. Uwe Flick, Ernst von Kardorff, and Ines Steinke, 589–600. Reinbek: Rowohlt.

———. 2002. Qualitative Interviews. Ein Überblick. In *Qualitative Forschung Ein Handbuch*, ed. Uwe Flick, Ernst von Kardorff, and Ines Steinke, 349–360. Reinbek: Rowohlt.

Houtkoop-Steenstra, Hanneke. 2000. *Interaction and the standardized survey interview. The living questionnaire*. Cambridge: Cambridge University Press.

Huberman, A. Michael, and Matthew B. Miles. 1998. Data management and analysis methods. In *Collecting and interpreting qualitative materials*, ed. Norman K. Denzin and Yvonna Lincoln, 179–211. Thousand Oaks: Sage.

Hundsnurscher, Franz. 1975. Semantik der Fragen. *Zeitschrift für Germanistische Linguistik* 3: 1–14.

Hymes, Dell. 1974. *Foundations in sociolinguistic. An ethnographic approach*. Philadelphia: University of Pennsylvania Press.

Isenberg, Horst. 1978. Probleme der Texttypologie. Variation und Determination von Texttypen. *WZ Leipzig* 27: 565–579.

Jahoda, Marie, Paul Felix Lazarsfeld, and Hans Zeisel. 2001. *Die Arbeitslosen von Marienthal. Ein soziographischer Versuch über die Wirkungen langandauernder Arbeitslosigkeit*. Frankfurt a. M.: Suhrkamp (Erstveröffentlichung 1933).

James, Nalita, and Hugh Busher. 2009. *Online interviewing*. Los Angeles: Sage.

Jones, Rodney. 2004. The Problem of Context in Computer-Mediated Communication. In *Discourse and technology: Multimodal discourse analysis*, ed. Philip Le Vine and Ron Scollon, 20–33. Washington, D.C.: Georgetown UP.

Jung, Thomas, and Stefan Müller-Dohm, eds. 1993. *"Wirklichkeit" im Deutungsprozeß. Verstehen und Methoden in den Kultur- und Sozialwissenschaften*. Suhrkamp: Frankfurt a. M.

Kallmeyer, Werner, and Fritz Schütze. 1977. Zur Konstitution von Kommunikationsschemata der Sachverhaltsdarstellung. In *Gesprächsanalysen*, ed. Dirk Wegner, 159–274. Hamburg: Buske.

Kelle, Udo. 2002. Computergestützte Analyse qualitativer Daten. In *Qualitative Forschung. Ein Handbuch*, ed. Uwe Flick, Ernst von Kardorff, and Ines Steinke, 485–503. Reinbek: Rowohlt.

Kelle, Udo, and Susann Kluge. 1999. *Vom Einzelfall zum Typus. Fallvergleich und Fallkontrastierung in der qualitativen Sozialforschung*. Opladen: Leske + Budrich.

Kelle, Udo, Susann Kluge, and Gerald Prein. 1993. *Strategien der Geltungssicherung in der qualitativen Sozialforschung. Zur Validitätsproblematik im interpretativen Paradigma*. Arbeitspapier Nr. 24 des Sonderforschungsbereichs 186, Bremen.

Keller, Reiner. 1997. Diskursanalyse. In *Sozialwissenschaftliche Hermeneutik*, ed. Ronald Hitzler and Anne Honer, 309–333. Opladen: Leske + Budrich.

———. 2007. Diskurse und Dispositive analysieren. Die Wissenssoziologische Diskursanalyse als Beitrag zu einer wissensanalytischen Profilierung der Diskursforschung. *Forum Qualitative Sozialforschung* 8 (2): Art. 19. http://nbn-resolving.de/urn:nbn:de:0114-fqs0702198.

———. 2012. *Das interpretative Paradigma. Eine Einführung*. Wiesbaden: Springer VS.

———. 2014. Wissenssoziologische Diskursforschung und Deutungsmusteranalyse. In *Wissen, Methode, Geschlecht: Erfassen des fraglos Gegebenen*, ed. Cornelia Behnke, Diana Lengersdorf, and Sylka Scholz, 143–159. Wiesbaden: Springer VS.

Kienpointner, Manfred. 1992. *Alltagslogik. Struktur und Funktion von Argumentationsmustern*. Stuttgart: Frommann-Holzboog.

King, Nigel, and Christine Horrock. 2010. *Interviews in qualitative research*. Los Angeles: Sage.

Kirk, Jerome, and Marc L. Miller. 1986. *Reliability and validity in qualitative research*. Beverly Hills: Sage.

Kleemann, Frank, Uwe Krähnke, and Ingo Matuschek. 2009. *Interpretative Sozialforschung. Eine praxisorientierte Einführung*. Wiesbaden: VS Verlag.

Klinke, Wolfgang. 1975. Wie heißt die Antwort auf diese Frage? In *Linguistisches Kolloquium 10*, ed. Heinrich Weber and Harald Weydt, 123–132. Berlin: De Gruyter.

Kluge, Susann. 1999. *Empirisch begründete Typenbildung. Zur Konstruktion von Typen und Typologien in der qualitativen Sozialforschung*. Opladen: Leske + Budrich.

Knoblauch, Hubert. 2005. *Wissenssoziologie*. Konstanz: UVK.

Kohli, Martin. 1978. "Offenes" und "geschlossenes" Interview: Neue Argumente zu einer alten Kontroverse. *Soziale Welt* 29: 1–23.

Konderding, Klaus-Peter. 2008. Diskurse, Topik, Deutungsmuster. Zur Komplementarität. Konvergenz und Explikation sprach-, kultur- und sozialwissenschaftlicher Zugänge zur Diskursanalyse auf der Grundlage kollektiven Wissens. In *Methoden der Diskurslinguistik. Sprachwissenschaftliche Zugänge zur transtextuellen Ebene*, ed. Ingo H. Warnke and Jürgen Spitzmüller, 117–150. Berlin: De Gruyter.

Kruse, Jan. 2014. *Qualitative Interviewforschung. Ein integrativer Ansatz*. Weinheim: Beltz.

Kuckartz, Udo. 2010. Typenbildung. In *Handbuch Qualitative Forschung in der Psychologie*, ed. Günter Mey and Katja Mruck, 553–568. Wiesbaden: Springer.

Kunze, Katharina. 2011. *Professionalisierung als biographisches Projekt: Professionelle Deutungsmuster und biographische Ressourcen von Waldorflehrerinnen und – Lehrern*. Wiesbaden: VS Verlag.

Kvale, Steinar. 1995. Validierung. Von der Beobachtung zu Kommunikation und Handeln. In *Handbuch Qualitative Sozialforschung*, ed. Uwe Flick, 427–431. München: Beltz & PVU.

———. 2007. *Doing interviews*. Los Angeles: Sage.

Kvale, Steinar, and Svend Brinkmann. 2009. *Interviews. Learning the craft of qualitative research interviewing*. 2nd ed. Los Angeles: Sage.

Labov, William. 1972. The transformation of experience in narrative syntax. In *Language in the inner city. Studies in the Black English vernacular*, ed. William Labov, 354–396. Philadelphia: University of Philadelphia Press.

Labov, William, and Joshua Waletzky. 1967. Narrative analysis: Oral versions of personal experience. In *Essays on the verbal and visual arts*, ed. June Helms, 12–44. Seattle: University of Washington Press.

Ladanyi, Peter. 1965. Zur logischen Analyse der Fragesätze. *Acta Linguistica Academiae Scientiarum Hungaricae* 15: 37–66.

Lamnek, Siegfried. 1988. *Qualitative Sozialforschung, Bd. 1., Methodologie*. München: Psychologie Verlags Union.

Lang, Jürgen. 1993. Frage und Fragehandlung. *Romanistisches Jahrbuch* 44: 43–56.

Lather, Patti. 1993. Fertile Obsession: Validity after Poststructuralism. *The Sociological Quarterly* 34 (4): 673–693.

Leibfried, Stephan, and Wolfgang Voges, eds. 1992. *Armut im modernen Wohlfahrtsstaat Sonderheft 32 der Kölner Zeitschrift für Soziologie und Sozialpsychologie*. Opladen: Westdeutscher Verlag.

Lewis, John A. 2009. Redefining qualitative methods: Believability in the fifth moment. *International Journal of Qualitative Methods* 8 (2): 1–14.

Lincoln, Yvonna S., and Egon G. Guba. 1985. *Naturalistic inquiry*. London: Sage.

Lucius-Hoene, Gabriele, and Arnulf Deppermann. 2004. *Rekonstruktion narrativer Identität. Ein Arbeitsbuch zur Analyse narrativer Interviews*. Wiesbaden: Verlag für Sozialwissenschaften.

Lüdemann, Christian. 1992. Deutungsmuster und das Modell rationalen Handelns: Eine Anwendung auf Deutungsmuster männlicher Sexualität. In *Analyse sozialer Deutungsmuster. Beiträge zur empirischen Wissenssoziologie*, ed. Michael Meuser and Reinhold Sackmann, 115–138. Pfaffenweiler: Centaurus.

Lüders, Christian. 1991. Deutungsmusteranalyse. Annäherung an ein risikoreiches Konzept. In *Qualitativ-empirische Sozialforschung: Konzepte, Methoden, Analysen*, ed. Detlef Garz and Klaus Kraimer, 377–408. Opladen: Westdeutscher Verlag.

———. 2003. Gütekriterien. In *Hauptbegriffe Qualitativer Sozialforschung*, ed. Ralf Bohnsack, Winfried Marotzki, and Michael Meuser, 80–82. Opladen: Leske + Budrich.

Lüders, Christian, and Michael Meuser. 1997. Deutungsmusteranalyse. In *Sozialwissenschaftliche Hermeneutik*, ed. Ronald Hitzler and Anne Honer, 57–79. Opladen: Leske + Budrich.

Madill, Anna, Abbie Jordan, and Caroline Shirley. 2000. Objectivity and reliability in qualitative analysis. Realist, contextualist and radical constructionist epistemologies. *British Journal of Psychology* 91: 1–20.

Malinowski, Bronislaw. 1923. The problem of meaning in primitive languages. Anhang. In *The meaning of meaning. A study of the influence of language upon thought and of the science of symbolism*, ed. Charles K. Ogden and Ivor A. Richards, 296–336. New York: Harcourt.

Mangold, Werner. 1973. Gruppendiskussionen. In *Handbuch der empirischen Sozialforschung, Bd. 2*, ed. René König. Stuttgart: Enke.

Mann, Chris, and Fiona Stewart. 2000. *Internet communication and qualitative research online. A handbook for researching online*. London: Sage.

Mannheim, Karl. 1980. *Strukturen des Denkens*. Hrsg. von David Kettler, Volker Meja, Nico Stehr. Frankfurt/M.: Suhrkamp.

Markova, Hristina. 2017. Neoliberale Hochschulpolitik? Deutungsmuster hochschulpolitischer Eliten am Beispiel der Exzellenzinitiative. In *Kapitalismus als Lebensform. Deutungsmuster, Legitimation und Kritik in der Marktgesellschaft*, ed. Patrick Sachweh and Sascha Münnich, 141–162. Wiesbaden: Springer VS.

Matthiesen, Ulf. 1989. "Bourdieu" und "Konopka". Imaginäres Rendezvous zwischen Habituskonstruktion und Deutungsmusterrekonstruktion. In *Klassenlage, Lebensstil und kulturelle Praxis. Theoretische und empirische Beiträge zur Auseinandersetzung mit Pierre Bourdieus Klassentheorie*, ed. Klaus Eder, 221–229. Frankfurt a. M.: Suhrkamp.

Maxwell, Joseph A. 1992. Understanding and validity in qualitative research. *Harvard Educational Review* 62 (3): 279–300.

Maynard, Douglas W., Hanneke Houtkoop-Steenstra, Nora C. Schaeffer, and Johannes van der Zouwen, eds. 2002. *Standardization as tacit knowledge. Interaction and practice in the survey interview*. New York: Wiley.

Mayntz, Renate. 1985. Über den begrenzten Nutzen methodologischer Regeln in der Sozialforschung. In *Entzauberte Wissenschaft. Zur Relativität und Geltung soziologischer Forschung. Soziale Welt, Sonderband 300*, ed. Wolfgang Bonß and Heinz Hartmann. Göttingen: Otto Schwarz.

Mayring, Philipp. 1983. *Qualitative Inhaltsanalyse. Grundlagen und Techniken*. Weinheim: Beltz.

McHoul, A.W. 1987. Why there are no guaranties for interrogators. *Journal of Pragmatics* 11: 455–471.

Mensching, Anja. 2010. "Goldfasan" versus "Kollege vom höheren Dienst". Zur Rekonstruktion gelebter Hierarchiebeziehungen in der Polizei. In *Das Gruppendiskussionsverfahren in der Forschungspraxis*, ed. Ralf Bohnsack, Aglaja Przyborski, and Burkhard Schäffer, 153–167. Opladen: B. Budrich.

Merkens, Hans. 1997. Stichproben bei qualitativen Studien. In *Handbuch Qualitative Forschungsmethoden in der Erziehungswissenschaft*, ed. Barbara Friebertshäuser and Annedore Prengel, 97–106. Weinheim: Juventa.

Metcalfe, Mike. 2005. Generalisation: Learning across epistemologies. *Forum qualitative Sozialforschung* 6 (1): Art. 17. http://nbn-resolving.de/urn:nbn:de:0114-fqs0501175.

Meulemann, Heiner. 1993. Befragung und Interview über soziale und soziologische Situationen der Informationssuche. *Soziale Welt* 44 (1): 98–119.

Meuser, Michael. 1992. "Das kann doch nicht wahr sein". Positive Diskriminierung und Gerechtigkeit. In *Analyse sozialer Deutungsmuster: Beiträge zur empirischen Wissenssoziologie*, ed. Michael Meuser and Rainer Sackmann, 89–102. Pfaffenweiler: Centaurus.

———. 2010. *Geschlecht und Männlichkeit. Soziologische Theorie und kulturelle Deutungsmuster*. 3rd ed. Wiesbaden: VS Verlag.

Meuser, Michael, and Ulrike Nagel. 1991. ExpertInneninterviews – Vielfach erprobt wenig bedacht Ein Beitrag zur qualitativen Methodendiskussion. In *Qualitativ-empirische Sozialforschung. Konzepte, Methoden, Analysen*, ed. Garz Detlef and Kraimer Klaus, 441–471. Opladen: WV.

Meuser, Michael, and Rainer Sackmann. 1992. Zur Einführung: Deutungsmusteransatz und empirische Wissenssoziologie. In *Analyse sozialer Deutungsmuster: Beiträge zur empirischen Wissenssoziologie*, ed. Michael Meuser and Rainer Sackmann, 9–37. Pfaffenweiler: Centaurus.

Mey, Günter. 2000. Erzählungen in qualitativen Interviews: Konzepte, Probleme, soziale Konstruktion. *Sozialer Sinn* 1: 135–151.

Mey, Inger. 2001. The CA/CDA controversy. *Journal of Pragmatics* 33: 609–615.

Mill, John Stuart. 1843. *A system of logic, ratiocinative and inductive. Being a connected view of the principles of evidence and the methods of scientific investigation*. London: Longmans, Green.

Miller, Gale, and Robert Dingwall, eds. 1997. *Context and method in qualitative research*. London: Sage.

Mishler, Elliot G. 1986. *Research interviewing. Context and narrative*. Cambridge: Harvard University Press.

———. 1990. Validation in inquiry-guided research. The role of expemplars in narrative studies. *Harvard Educational Review* 60 (4): 415–442.

———. 2005. Commentary 3. *Qualitative Research in Psychology* 2 (4): 315–318.

Misoch, Sabina. 2015. *Qualitative interviews*. München: Oldenbourg.

Mommsen, Wolfgang. 1974. "Verstehen" und "Idealtypus". Zur Methodologie einer historischen Sozialwissenschaft. In *Gesellschaft, Politik und Geschichte*, ed. Wolfgang Mommsen and Max Weber. Frankfurt a. M.: Suhrkamp.

Mühlfeld, Claus, Paul Windolf, Norbert Lampert, and Heidi Krüger. 1981. Auswertungsprobleme offener interviews. *Soziale Welt* 32 (3): 325–352.

Müller, Matthias. 2013. Deutungsmusteranalyse in der soziologischen Sozialpolitikforschung. Überlegungen zu einem qualitativen Forschungsansatz. *Zeitschrift für Qualitative Forschung* 14 (2): 295–310.

———. 2014. Dynamik sozialpolitischer Innovationen. Deutungsmuster zum Bedingungslosen Grundeinkommen. *Zeitschrift für Sozialreform* 60 (3): 295–320.

Murphy, E., R. Dingwall, D. Greatbatch, S. Parker, and P. Watson. 1998. Qualitative research methods in health technology assessment. A review of the literature. *Health Technology Assessment* 2 (16): 1–260.

Nentwig-Gesemann, Iris. 2001. Die Typenbildung der dokumentarischen Methode. In *Die dokumentarische Methode und ihre Forschungspraxis. Grundlagen qualitativer Sozialforschung*, ed. Ralf Bohnsack, Iris Nentwig-Gesemann, and Arnd-Michael Nohl, 275–300. Opladen: Leske + Budrich.

Neuendorff, Hartmut, and Charles Sabel. 1978. Zur relativen Autonomie der Deutungsmuster. In *Materialien aus der soziologischen Forschung. Verhandlungen des 18. Deutschen Soziologentages vom 28. September bis 1. Oktober 1976 in Bielefeld*, ed. Karl-Martin Bolte, 842–863. Darmstadt: Luchterhand.

Nohl, Arnd-Michael. 2013. *Relationale Typenbildung und Mehrebenenvergleich. Neue Wege der dokumentarischen Methode*. Wiesbaden: Springer VS.

Oakley, Ann. 1981. Interviewing women: A contradiction in terms. In *Doing feminist research*, ed. Helen Roberts, 30–61. London: Routledge.

Oevermann, Ulrich. 1973. *Zur Analyse der Struktur von sozialen Deutungsmustern (unv. Ms.)*. [später (2001) in Sozialer Sinn 2001 (1), S. 3–34. abgedruckt].

———. 1997. *Thesen zur Methodik der werkimmanenten Interpretation vom Standpunkt der objektiven Hermeneutik*. Vorgelegt zur 4. Arbeitstagung der Arbeitsgemeinschaft objektive Hermeneutik e. V. "Immanenz oder Kontextabhängigkeit? Zur Methodik der Analyse von Werken und ästhetischen Ereignissen" am 26./27. April 1997 in Frankfurt a. M. (Ms.).

———. 2001. Die Struktur sozialer Deutungsmuster. *Versuch einer Aktualisierung. Sozialer Sinn* 2001 (1): 35–81.

———. 2002. *Klinische Soziologie auf der Basis der Methodologie der objektiven Hermeneutik – Manifest der objektiv hermeneutischen Sozialforschung*. Frankfurt a. M.: Institut für hermeneutische Sozial- und Kulturforschung.

Oevermann, Ulrich, Tilman Allert, Elisabeth Konau, and Jürgen Krambeck. 1979. Die Methodologie einer "objektiven Hermeneutik" und ihre allgemeine forschungslogische Bedeutung in den Sozialwissenschaften. In *Interpretative Verfahren in den Sozial- und Textwissenschaften*, ed. Hans-Georg Soeffner, 352–434. Stuttgart: Metzlersche Verlagsbuchhandlung.

Onwuegbuzie, Anthony J., and Nancy L. Leech. 2007. Validity and qualitative research: An oxymoron? *Quality & Quantity* 41 (2): 233–249.

Opielka, Michael, Müller, Matthias, Bendixen, Tim, and Kreft, Jesco. 2009. *Grundeinkommen und Werteorientierungen. Eine empirische Analyse*. Wiesbaden: VS-Verlag.

Pareto, Vilfredo. 1955. *Allgemeine Soziologie. Ausgewählt und übersetzt von C. Brinkmann*. Tübingen: Mohr (Erstveröffentlichung 1916).

Patton, Michael Quinn. 1990. *Qualitative evaluation and research methods*. Newbury Park: Sage.

Pfister, Bernhard. 1928. *Die Entwicklung zum Idealtypus*. Eine methodologische Untersuchung über das Verhältnis von Theorie und Geschichte bei Menger, Schmoller und Max Weber. Tübingen, Mohr.

Plaß, Christine, and Michael Schetsche. 2001. Grundzüge einer wissenssoziologischen Theorie sozialer Deutungsmuster. *Sozialer Sinn* 3: 511–536.

Popitz, Heinrich, Hans-Paul Bahrdt, Ernst-August Jüres, and Hanno Kesting. 1957. *Das Gesellschaftsbild des Arbeiters. Soziologische Untersuchungen in der Hüttenindustrie*. Tübingen: Mohr.

Potter, Jonathan, and Alexa Hepburn. 2005. Qualitative interviews in psychology: Problems and possibilities. *Qualitative Research in Psychology* 2 (4): 281–307.

Przyborski, Aglaja. 2004. *Gesprächsanalyse und dokumentarische Methode. Qualitative Auswertung von Gesprächen, Gruppendiskussionen und anderen Diskursen*. Wiesbaden: VS Verlag.

Przyborski, Aglaja, and Julia Riegler. 2010. Gruppendiskussion und Fokusgruppe. In *Handbuch qualitative Forschung in der Psychologie*, ed. Günter Mey and Katja Mruck, 436–448. Wiesbaden: Springer.

Przyborski, Aglaja, and Monika Wohlrab-Sahr. 2014. *Qualitative Sozialforschung. Ein Arbeitsbuch*. 2nd ed. München: Oldenbourg.

Quasthoff, Uta. 1979. Eine interaktive Funktion von Erzählungen. In *Interpretative Verfahren in den Sozial- und Textwissenschaften*, ed. Hans-Georg Soeffner, 104–126. Stuttgart: Metzler.

Reichertz, Jo. 1988. Verstehende Soziologie ohne Subjekt? Die objektive Hermeneutik als Metaphysik der Strukturen. *Kölner Zeitschrift für Soziologie und Sozialpsychologie* 40: 207–221.

———. 1997. Plädoyer für das Ende einer Methodologiedebatte bis zur letzten Konsequenz. In *Beobachtung verstehen, Verstehen beobachten. Perspektiven einer konstruktivistischen Hermeneutik*, ed. Tillman Sutter, 98–132. Opladen: Westdeutscher Verlag.

———. 1999. Über das Problem der Gültigkeit von Qualitativer Sozialforschung. In *Hermeneutische Wissenssoziologie. Standpunkte zur Theorie der Interpretation*, ed. Ronald Hitzler, Jo Reichertz, and Norbert Schröer, 319–346. Konstanz: UVK.

Reinders, Heinz. 2005. *Qualitative interviews mit Jugendlichen führen. Ein Leitfaden*. München: Oldenbourg.

Reisigl, Martin, and Ruth Wodak. 2009. The discourse-historical approach (DHA). In *Methods of critical discourse analysis*, ed. Ruth Wodak and Michael Meyer, 87–121. Los Angeles: Sage.

Richardson, Stephen A., Barbara Snell Dohrenwend, and David Klein. 1965. *Interviewing. Its forms and functions*. New York: Basic Books.

———. 1979. Die "Suggestivfrage". Erwartungen und Unterstellungen im interview. In *Qualitative Sozialforschung*, ed. Christel Hopf and Elmar Weingarten, 205–231. Stuttgart: Klett-Cotta (Erstveröffentlichung 1965).

Ricoeur, Paul. 1972. Der Text als Modell: Hermeneutisches Verstehen. In *Verstehende Soziologie*, ed. Walter L. Bühl, 252–283. München: Nymphenburger.

Rogers, Carl R. 1993. *Die klientenzentrierte Gesprächspsychotherapie*. Frankfurt a. M.: Fischer.

Rosenthal, Gabriele. 2008. *Interpretative Sozialforschung. Eine Einführung*. 2nd ed. Weinheim: Juventa.

Rost-Roth, Martina. 2006. *Nachfragen. Formen und Funktionen äußerungsbezogener Interrogationen*. Berlin: De Gruyter.

Roulston, Kathryn. 2014. *Reflective interviewing. A guide to theory and practice*. 2nd ed. London: Sage.

Rüling, Anneli. 2007. *Jenseits der Traditionalisierungsfallen. Wie Eltern sich Familien- und Erwerbsarbeit teilen*. Frankfurt/M.: Campus.

Sachweh, Patrick. 2010. *Deutungsmuster sozialer Ungleichheit. Wahrnehmung und Legitimation gesellschaftlicher Privilegierung und Benachteiligung*. Frankfurt a. M.: Campus.

Saldaña, Johnny. 2009. *The coding manual for qualitative researchers*. London: Sage.

Schatzman, Leonard, and Anselm Strauss. 1973. *Field research. Strategies for a natural sociology*. Englewood Cliffs: Prentice Hall.

Scheele, Brigitte, and Norbert Groeben. 1988. *Dialog-Konsens-Methoden zur Rekonstruktion Subjektiver Theorien*. Tübingen: Francke.

Schegloff, Emanuel A. 2002. Survey interviews as talk-in-interaction. In *Standardization as Tacit Knowledge. Interaction and Practice in the Survey Interview*, ed. Douglas W. Maynard et al., 151–157. New York: Wiley.

Schetsche, Michael. 2000. *Wissenssoziologie sozialer Probleme. Grundlegungen einer relativistischen Problemtheorie*. Wiesbaden: Westdeutscher Verlag.

Schiek, Daniela. 2014. Das schriftliche Interview in der qualitativen. *Zeitschrift für Soziologie* 43 (5): 379–395.

Schiek, Daniela, and Carsten G. Ullrich, eds. 2016. *Qualitative Online-Erhebungen. Voraussetzungen – Möglichkeiten – Grenzen*. Wiesbaden: Springer VS.

Scholz, Antonia. 2012. *Migrationspolitik zwischen moralischem Anspruch und strategischem Kalkül. Der Einfluss politischer Ideen in Deutschland und Frankreich*. Wiesbaden: Springer VS.

Schütz, Alfred, and Thomas Luckmann. 1979. *Strukturen der Lebenswelt, Bd. 2*. Frankfurt a. M.: Suhrkamp.

Schütze, Fritz. 1976a. Zur Hervorlockung und Analyse von Erzählungen thematisch relevanter Geschichten im Rahmen soziologischer Feldforschung – Dargestellt an einem Projekt zur Erforschung kommunaler Machtstrukturen. In *Kommunikative Sozialforschung*, ed. Arbeitsgruppe Bielefelder Soziologen, 159–260. München: Fink.

———. 1976b. *Zur soziologischen und linguistischen Analyse von Erzählungen. Internationales Jahrbuch für Wissens- und Religionssoziologie, Bd. 10*. Opladen: Westdeutscher Verlag.

———. 1977. *Die Technik des narrativen Interviews in Interaktionsfeldstudien – Dargestellt an einem Projekt zur Erforschung von kommunalen Machtstrukturen*. Universität Bielefeld, Fakultät für Soziologie, Arbeitsberichte und Forschungsmaterialien. Bielefeld, Manuskript Nr. 1.

———. 1983. Biographieforschung und narratives Interview. *Neue Praxis* 13: 283–293.

Schütze, Yvonne. 1986. *Die gute Mutter. Zur Geschichte des normativen Musters "Mutterliebe"*. Bielefeld: Kleine.

Schütze, Fritz. 1987. *Das narrative Interview in Interaktionsfeldstudien I*. Studienbrief der Fernuniversität Hagen.

Seale, Clive. 1999. *The quality of qualitative research*. London: Sage.

Searle, John R. 1971. *Sprechakte. Ein sprachphilosophischer Essay*. Frankfurt a. M.: Suhrkamp.

Seidman, Irving. 1998. *Interviewing as qualitative research. A guide for researchers in education and social Sciences*. New York: Teachers College.

Sikora, Jan. 2012. Fragen als Sprechakt (Zur kommunikativen Funktion von Fragen). *Studia Germanica Gedanensia* 27: 101–111.

Silverman, David. 1973. Interview talk: Bringing off a research instrument. *Sociology* 7 (1): 31–48.

———. 1997. The logics of qualitative research. In *Context and method in qualitative research*, ed. Gale Miller and Robert Dingwall, 12–25. London: Sage.

Soeffner, Hans-Georg, ed. 1979. *Interpretative Verfahren in den Sozial- und Textwissenschaften*. Stuttgart: Metzler.

———. 1989. *Auslegung des Alltags – Der Alltag der Auslegung*. Frankfurt a. M.: Suhrkamp.

Steinke, Ines. 1999. *Kriterien qualitativer Forschung. Ansätze zur Bewertung qualitativ-empirischer Sozialforschung*. Weinheim: Juventa.

Stenvoll, Dag, and Peter Svensson. 2011. Contestable contexts. The transparent anchoring of contextualization in text-as-data. *Qualitative Research* 11 (5): 570–586.

Strauss, Anselm L. 1991. *Grundlagen Qualitativer Sozialforschung*. München: Fink.

Strauss, Anselm L., and Jeanette Corbin. 1990. *Basics of qualitative research*. London: Sage.

Streckeisen, Ursula, Denis Hänzi, and Andrea Hungerbühler. 2007. *Fördern und Auslesen. Deutungsmuster von Lehrpersonen zu einem beruflichen Dilemma*. Wiesbaden: VS Verlag.

Strübing, Jörg. 2008. *Grounded Theory. Zur sozialtheoretischen und epistemologischen Fundierung des Verfahrens der empirisch begründeten Theoriebildung*. 2nd ed. Wiesbaden: VS.

Strübing, Jörg, Stefan Hirschauer, Ruth Ayaß, Uwe Krähnke, and Thomas Scheffer. 2018. Gütekriterien qualitativer Sozialforschung. Ein Diskussionsanstoß. *Zeitschrift für Soziologie* 47 (2): 83–100.

Talmy, Steven. 2011. The interview as collaborative achievement. Interaction, identity, and ideology in a speech event. *Applied Linguistics* 32 (1): 25–42.

Tashakkori, Abbas, and Charles Teddlie, eds. 2003. *Handbook for mixed methods in social and behavioral research*. Thousand Oaks: Sage.

ten Have, Paul. 2004. *Understanding qualitative research and ethnomethodology*. London: Sage.

———. 2007. *Doing conversation analysis. A practical guide*. 2nd ed. London: Sage.

Terhart, Ewald. 1995. Kontrolle von Interpretationen: Validierungsprobleme. In *Bilanz qualitativer Forschung, Bd. I. Grundlagen qualitativer Forschung*, ed. Eckard König and Peter Zeller, 373–397. Weinheim: Deutscher Studien Verlag.

Thomssen, Wilke. 1980. Deutungsmuster – Eine Kategorie der Analyse gesellschaftlichen Bewußtseins. In *Handbuch zur Soziologie der Weiterbildung*, ed. Ansgar Weymann, 358–373. Darmstadt: Luchterhand.

Ullrich, Carsten G. 1999. Deutungsmusteranalyse und diskursives Interview. *Zeitschrift für Soziologie* 28 (6): 429–447.

———. 2000. *Solidarität im Sozialversicherungsstaat. Die Akzeptanz des Solidarprinzips in der Gesetzlichen Krankenversicherung*. Frankfurt a. M.: Campus.

———. 2008. *Die Akzeptanz des Wohlfahrtsstaates. Präferenzen, Konflikte, Deutungsmuster*. Wiesbaden: VS Verlag.

Ullrich, Carsten G., and Daniela Schiek. 2014. Gruppendiskussionen in Internetforen. Zur Methodologie eines neuen qualitativen Erhebungsinstruments. *Kölner Zeitschrift für Soziologie und Sozialpsychologie* 66 (3): 459–474.

van Oorschot, Wim, and Halman Loek. 2000. Blame or fate? Individual or social? An international comparison of popular explanations of poverty. *European Societies* 2: 1–28.

Vandeweghe, Willy. 1977. Fragen und ihre Funktionen. Versuch einer Typologie auf pragmatischer Basis. In *Semantik und Pragmatik. Akten des 11. Linguistischen Kolloquiums Aachen 1976, Bd. 2*, ed. Konrad Sprengel et al., 277–286. Tübingen: Niemeyer.

von Kardorff, Ernst. 1995. Qualitative Sozialforschung – Versuch einer Standortbestimmung. In *Handbuch qualitativer Sozialforschung*, ed. Uwe Flick et al., 3–8. München: Psychologie Verlagsunion.

von Unger, Hella. 2014. Forschungsethik in der qualitativen Forschung: Grundsätze, Debatten und offene Fragen. In *Forschungsethik in der qualitativen Forschung. Reflexivität, Perspektiven, Positionen*, ed. Petra Narimani and Rosaline M'Bayo, 15–40. Wiesbaden: Springer VS.

Wagner, Klaus R. 2001. *Pragmatik der deutschen Sprache*. Frankfurt a. M.: Lang.

Wagner, Hans. 2008. *Qualitative Methoden in der Kommunikationswissenschaft. Ein Lehr- und Studienbuch. Unter Mitarbeit von Philomen Schönhagen, Ute Nawratil und Heinz Starkulla*. München: Fischer.

Waldenfels, Bernhard. 1994. *Antwortregister*. Frankfurt a. M.: Suhrkamp.

Walther, Jürgen. 1975. Zu Logik von Frage und Antwort. In *Linguistisches Kolloquium 10*, ed. Heinrich Weber and Harald Weydt, 133–141. Berlin: De Gruyter.

Weber, Max. 1985. *Wirtschaft und Gesellschaft. Grundriß der verstehenden Soziologie*. Tübingen: Mohr (Erstveröffentlichung 1921).

———. 1995. *Gesammelte Aufsätze zur Wissenschaftslehre*. Tübingen: Mohr (Erstveröffentlichung 1922).

Weigelt, Linda. 2010. *Berührungen und Schule. Deutungsmuster von Lehrkräften. Eine Studie zum Sportunterricht*. Wiesbaden: VS Verlag.

Wernet, Andreas. 2000. *Einführung in die Interpretationstechnik der Objektiven Hermeneutik*. Opladen: Leske + Budrich.

Whyte, William Foote. 1996. *Die Street Corner Society. Die Sozialstruktur eines Italienerviertels*. Berlin: De Gruyter (Erstveröffentlichung 1943).

Williams, Malcolm. 2000. Interpretivism and generalisation. *Sociology* 34 (2): 209–224.

Wilson, Thomas P. 1982. Qualitative "oder" quantitative Methoden in der Sozialforschung. *Kölner Zeitschrift für Soziologie und Sozialpsychologie* 34: 469–486.

Witzel, Andreas. 1985. Das problemzentrierte Interview. In *Qualitative Forschung in der Psychologie*, ed. Gerd Jüttemann, 227–255. Weinheim: Beltz.

Witzel, Andreas, and Herwig Reiter. 2012. *The problem-centred interview*. London: Sage.

Wolde, Anja. 2007. *Väter im Aufbruch? Deutungsmuster von Väterlichkeit und Männlichkeit im Kontext von Väterinitiativen*. Wiesbaden: VS Verlag.

Wolff, Stephan. 1986. Das Gespräch als Handlungsinstrument. Konversationsanalytische Aspekte sozialer Arbeit. *Kölner Zeitschrift für Soziologie und Sozialpsychologie* 38: 55–84.

Wolff, Stephan, and Claudia Puchta. 2007. *Realitäten zur Ansicht. Die Gruppendiskussion als Ort der Datenproduktion*. Stuttgart: Lucius & Lucius.

Wunderlich, Dieter. 1976. *Studien zur Sprechakttheorie*. Frankfurt a. M.: Suhrkamp.

Yang, Young-Sook. 2003. *Aspekte des Fragens. Frageäußerungen, Fragesequenzen, Frageverben*. Berlin: De Gruyter.

Index

C. G. Ullrich, *The Discursive Interview*,
https://doi.org/10.1007/978-3-658-38477-7

The manufacturer's authorised representative in the EU is Springer Nature Customer Service Centre GmbH, Europaplatz 3, 69115 Heidelberg, Germany. If you have any concerns regarding our products, please contact ProductSafety@springernature.com

Printed and bound by CPI Group (UK) Ltd, Croydon, CR0 4YY
30/09/2026
02225450-0001